4721800021806

D0548035

50

IDEAS YOU REALLY NEED TO KNOW

ASTRONOMY

GILES SPARROW

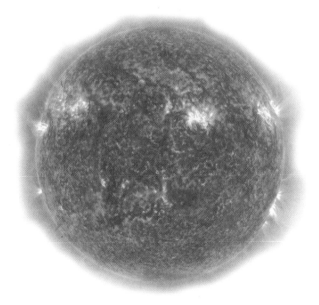

Quercus

Contents

Introduction

Considering how rarely the behaviour of objects in the night sky seems to have a direct impact on human lives, it might seem strange that astronomy can lay claim to being the oldest of the sciences. In fact, the roots of astronomy pre-date recorded history – the oldest known star map was painted onto the walls of a cave at Lascaux, France in the midst of the last Ice Age some 17,300 years ago. At first glance, it is simply a beautiful representation of a charging bull, but closer inspection reveals a cluster of dots behind the animal's hump: an unmistakeable representation of the Pleiades star cluster in the modern constellation of Taurus.

For the ancients, the motions of Sun, Moon and stars really did have an vital connection to events on Earth: technology may have diminished our exposure to the changing seasons, but for our ancestors, they were a matter of life and death. In the modern age, astronomy wields its influence in other ways, often through the scientific innovation it inspires (as the CCD camera in your smartphone attests). But perhaps the true fascination of astronomy in our bewildered modern times lies in the fact that it touches the mysteries of the infinite, and comes closer than any other science to explaining where we come from.

This book is a celebration of astronomy's greatest ideas, and the brilliant, insightful and sometimes iconoclastic minds that helped shape them. Across fifty topics, I hope to encompass everything from the varied planets and other worlds on our celestial doorstep, through the lives and deaths of stars, to the structure and origins of the Universe itself. Some of the theories discussed date back for centuries, while others are startlingly modern, and some are still in the process of formation – one of the great beauties of astronomy as a science is that, like the Universe itself, it never stands still. Inevitably, my selection of topics is a personal one, shaped by my own interests and discussions with numerous working astronomers, but I hope there is something here to fascinate, and perhaps even inspire, everyone.

Giles Sparrow

01 Our place in the Universe

The story of astronomy is one in which our understanding of our place in the Universe advances, while our significance within the cosmos gradually diminishes. Once at the centre of creation, our world is now seen as a speck in the vastness of the cosmos.

Humanity has been obsessed with the stars throughout history, not only telling tales about them and filling them with significance, but also using them for practical purposes such as keeping track of time. The ancient Egyptians predicted the Nile flood season was approaching when Sirius, the brightest star in the sky, rose shortly before dawn. But another important strand of ancient thought, astrology, produced the first attempts to model our place in the cosmos.

Ancient astrologers were driven by the idea that the heavens were a mirror of the Earth: the movements of the Sun, Moon and wandering planets among the fixed star patterns called constellations did not necessarily influence events on Earth, but they did reflect them. Thus, if a great famine struck when Mars and Jupiter were in conjunction (close to each other in the sky) in Taurus, then you might anticipate a similar event when those planets neared alignment in that constellation once again. What was more, the movements of the planets were not entirely unpredictable, so if you could forecast their movements, you might be able to foretell future events on Earth.

TIMELINE

*c.*150 CE	1543	1608
Ptolemy's *Almagest* cements the classical view of an Earth-centred, geocentric Universe	Copernicus publishes his case for a Sun-centred, heliocentric Universe	Kepler models orbits as ellipses, rather than circles, finally explaining the motions of the planets

THE GEOCENTRIC UNIVERSE

The great challenge, then, was to develop a sufficiently accurate model of planetary motions. Most ancient astronomers were hamstrung by the common-sense idea that the Earth is fixed in space (after all, we do not feel its motion). With no inkling of the scale of the cosmos, they assumed that the Moon, Sun, planets and stars all followed circular paths around it at varying speeds, in such a way as to produce the apparent motions seen in the sky (see box, see page 6).

> **... THAT VAST UNIVERSE IN WHICH WE ARE IMBEDDED LIKE A GRAIN OF SAND IN A COSMIC OCEAN.**
> Carl Sagan

Unfortunately, this geocentric (Earth-centred) model, despite its appealing simplicity, did not make accurate predictions. Planets shifted rapidly from their predicted paths through the sky, and astronomers added various fudges to correct for this. The model reached its pinnacle in the 2nd century CE through the work of Greek–Egyptian astronomer Ptolemy of Alexandria. His great work, *Almagest*, envisioned planets moving on circular paths called epicycles whose centres orbited the Earth in turn. Endorsed by both the Roman Empire and its Christian and Muslim successors, Ptolemy's model reigned supreme for more than a millennium. Contemporary astronomers largely concerned themselves with refining measurements of planetary movements in order, they hoped, to tweak the model's various parameters and improve its predictions.

THE SUN AT THE CENTRE

With the dawn of the European Renaissance, the long-held view that ancient wisdom was unimpeachable began to founder among thinkers in a number of fields, and some astronomers began to wonder if Ptolemy's geocentric model was fundamentally flawed. In 1514, Polish priest Nicolaus Copernicus circulated a small book arguing that the observed motions of the heavens might be better explained by a Sun-centred, or heliocentric, model. In this conception, the Earth

1781
William Herschel makes the first map of the Milky Way, showing our galaxy as a flattened plane of stars

1924
Edwin Hubble shows that spiral nebulae are independent galaxies millions of light years beyond our own

1929
Hubble shows that the Universe is expanding – the root of the Big Bang theory

Planetary motions

Planets in Earth's skies are broadly divided into two groups – the 'inferior' planets Mercury and Venus make loops around the Sun's position in the sky, but never stray far from it and so always appear in the west after sunset, or the east before sunrise. In contrast, the 'superior' planets – Mars, Jupiter, Saturn, Uranus and Neptune – follow tracks that take them around the entire sky and can appear on the opposite side of the sky to the Sun. But their motion is complicated by retrograde loops, periods when they slow and temporarily reverse their eastward drift against the stars before eventually continuing on their way. Retrograde motion was the greatest challenge for geocentric models of the solar system, and Ptolemy explained it by placing the superior planets on orbits within orbits known as epicycles. In a heliocentric system, however, retrograde motion is fairly easy to explain as an effect of shifting points of view as the faster-moving Earth overtakes a superior planet.

is just one of several planets on circular paths around the Sun, and only the Moon actually orbits Earth (a theory that had in fact been proposed by several ancient Greek philosophers). Copernicus' idea began to gain ground with the posthumous publication of his masterwork *On the Revolutions of the Heavenly Spheres* in 1543, but his circular orbits caused problems of their own when it came to making accurate predictions. It was not until 1608, when German astronomer Johannes Kepler put forward a new model in which orbits were stretched ellipses, that the mystery of planetary motions was finally solved. Our world was relegated from its position at the heart of creation.

Soon, astronomers realized that the Copernican Revolution diminished our place in the Universe still further. If Earth was moving from side to side of a vast, sweeping orbit, then surely the parallax effect (the apparent shifting of nearby objects when seen from different points of view) should affect the positions of the stars? The fact that no parallax could be seen, even with new observing aids such as the telescope (see page 8), implied that the stars were unimaginably far away – not a sphere of lights around the solar system, but distant suns in their own right. What was more, telescopes revealed countless previously unseen stars and showed that the pale band of the Milky Way was made up of dense star clouds.

THE WIDER UNIVERSE

By the late 18th century, astronomers had begun to map the structure of our galaxy, the flattened plane of stars (later shown to be a disc, then a spiral – see page 136) that was thought to contain all of creation. At first, Earth was once again privileged by being placed near the galaxy's centre, and it was not until the 20th century that our solar system's true location – some

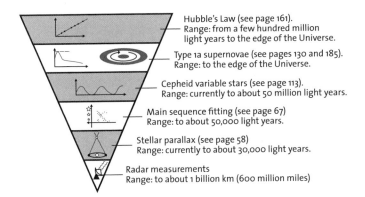

Hubble's Law (see page 161).
Range: from a few hundred million
light years to the edge of the Universe.

Type 1a supernovae (see pages 130 and 185).
Range: to the edge of the Universe.

Cepheid variable stars (see page 113).
Range: currently to about 50 million light years.

Main sequence fitting (see page 67)
Range: to about 50,000 light years.

Stellar parallax (see page 58)
Range: currently to about 30,000 light years.

Radar measurements
Range: to about 1 billion km (600 million miles)

A wide range of different techniques are used to measure distances of nearby and distant astronomical objects. Throughout the history of astronomy, establishing a rung on this distance ladder often reveals clues as to how objects on the next rung can be found.

26,000 light years out in a fairly unremarkable part of the Milky Way – was confirmed. By that time, breakthroughs in our understanding of the stars, including accurate measurements of their distances (see page 56) had shown that even our Sun was nothing special. In fact, it is a fairly dim yellow dwarf star, outshone by many of the 200 billion or more stars in our galaxy.

One final great shift in our cosmic perspective came in 1924, when US astronomer Edwin Hubble showed that the 'spiral nebulae' seen in various parts of the sky were in fact unimaginably distant star systems. The Milky Way, of which we are such an insignificant part, is itself just one among countless galaxies (see page 144) – perhaps as many as there are stars in our galaxy, scattered across an ever-expanding Universe (see page 160). And even this may not be the end of the story: there is growing evidence that our Universe itself may be just one of infinitely many in the unfathomable structure known as the multiverse (see page 196).

The condensed idea
Each new discovery diminishes our place in the Universe

02 Watching the skies

Telescopes have transformed the way we understand the Universe. Today's ground-based and orbiting observatories can peer to the very edge of space and resolve detail over vast distances, while other sophisticated instruments use invisible radiations to discover hidden aspects of the cosmos.

P rior to the invention of the telescope, the most important tools at an astronomer's disposal were astrolabes, quadrants and other devices used to measure the position of objects in the sky, and the angles between them. The unaided human eye placed natural limits on both the brightness of objects that could be seen and the amount of detail that could be distinguished. Then in 1608, a Dutch spectacle-maker called Hans Lippershey applied for a patent on an ingenious device using two lenses (a convex objective and a concave eyepiece) to create an image magnified by about three times. This was the first telescope.

A BETTER VIEW

News of the Dutch invention spread rapidly, with word reaching Galileo Galilei in Venice by June 1609. Working out the principles for himself, Galileo built several instruments, culminating in one with an unprecedented 33x magnification. In 1610, he made a number of important discoveries with this telescope, including the four bright satellites of Jupiter, spots on the Sun and the phases of Venus. These convinced him that the Copernican Sun-centred Universe was correct, and brought him into conflict with the conservative authorities of the Catholic Church.

TIMELINE

1609	1668	1870s
Galileo is one of the first people to point a telescope at the skies	Isaac Newton builds the first functional reflecting (mirror-based) telescope	William Huggins begins to use photography and spectroscopy through telescopes as a research tool

In 1611, Johannes Kepler worked out how, in principle, a telescope with two convex lenses could produce much higher magnifications, and by the mid-17th century this had become the most popular type of telescope, leading to many new discoveries. One particularly successful instrument-builder was Dutch scientist Christiaan Huygens, who used increasingly long telescopes to make discoveries including Saturn's moon Titan and the true shape of Saturn's rings (which Galileo had identified as a strange distortion).

However, the later 1600s saw an entirely new type of telescope rise to prominence. The reflector design used a curved primary mirror to gather and focus light, and a smaller secondary one to deflect it towards an eyepiece. The first practical telescope of this design was completed by Isaac Newton in 1668, and spawned many variants. Telescopes offer astronomers greater light grasp and improved resolving power. A telescope's objective lens or primary mirror offers a much larger collecting surface for faint starlight than the small diameter of a human pupil, and so telescopes are generally able to see much fainter objects. The magnifying power offered by the eyepiece, meanwhile, can allow us to resolve detail and separate closely spaced objects.

> **OUR KNOWLEDGE OF STARS AND INTERSTELLAR MATTER MUST BE BASED PRIMARILY ON THE ELECTROMAGNETIC RADIATION WHICH REACHES US.**
>
> Lyman Spitzer

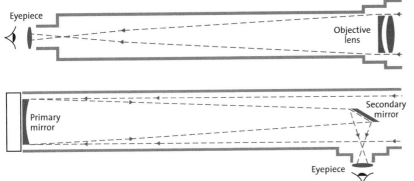

Eyepiece

Objective lens

Primary mirror

Secondary mirror

Eyepiece

Schematics of the two basic types of telescope. In a refractor (top) light collected by an objective lens is bent to a focus and then a magnified image is created by the eyepiece lens. In a Newtonian reflector (bottom), a curved primary mirror collects the light and bounces it back to a secondary mirror, which then deflects it to an eyepiece lens.

1957

Bernard Lovell builds the world's first large steerable radio telescope at Jodrell Bank, England

1979

First Multiple-Mirror Telescope is built at Mount Hopkins, Arizona

1990

The Hubble Space Telescope becomes the first large optical telescope in space

Pushing the limits

The latest generation of large astronomical telescopes make use of computer control and modern materials to create larger light-gathering surfaces than ever before. The largest single-mirror instruments are the twin 8.4-metre (27.5-ft) monsters of the Large Binocular Telescope at Arizona's Mount Graham International Observatory, with the four 8.2-metre (27-ft) mirrors of the European Southern Observatory's Very Large Telescope (VLT) in Chile not far behind. Both instruments use active optics – a web of computerized motors called actuators that support the mirror and counteract distortions caused by its own weight. Another system, called adaptive optics, measures the distortion of light from the target object as it passes through the atmosphere, and constantly adjusts the mirror to counter this, resulting in images whose sharpness can rival those from the Hubble Space Telescope.

Multiple-mirror telescopes can get even bigger. The Gran Telescopio Canarias on La Palma in the Canary Isles has 36 interlocking mirrors that give a surface equivalent to a single 10.4-metre (34-ft) mirror. Even more ambitious projects are being planned, with construction now underway in Chile on the European Extremely Large Telescope (E-ELT), whose enormous 39.3-metre (129-ft) primary mirror consists of 798 individual segments.

MODERN TELESCOPES

Both types of telescope have their pros and cons, but in general the practical problems of casting and mounting heavy convex lenses, plus the huge amounts of precious starlight they absorb, limits lens-based refractors to around 1 metre (40 in). The size of reflecting telescopes, meanwhile, stalled at around the 5-metre (200-in) level for much of the 20th century. However, new materials (mirrors made from interlocking honeycomb segments) and, above all, computerized control have allowed them to mushroom in size to 10 metres (400 inches) and larger (see box).

Of course, most modern telescopes are not built with the human eye in mind, and since the mid-19th century, photography has played an important role in astronomy. Photographs not only capture views for posterity, they also boost a telescope's light grasp still further. Provided the telescope is oriented correctly and rotated slowly to keep pace with the effects of Earth's spin, a long-exposure image can 'integrate' many hours' worth of distant starlight. Astronomical photography is now dominated by electronic CCDs, which can even track the precise number of photons striking an individual semiconductor pixel. Often, light from a distant object is passed through a spectroscope (a device with a finely scored diffraction grating that works in the same way as a prism), splitting it into a rainbow-like spectrum within which the intensity of specific colours can be measured as part of a spectroscopic investigation (see page 60).

INVISIBLE RADIATIONS

The visible light reaching Earth's surface from space is just one small part of an overall electromagnetic spectrum. Electromagnetic radiations consist of oscillating packets of waves called photons, and our eyes have evolved to see light because it happens to be one of the few bands of radiation that make it through Earth's atmosphere to the surface. Other forms of radiation include infrared ('heat radiation' with waves slightly longer than those of red light), and radio (with even longer waves). Infrared radiation from space tends to get swamped by the heat of our own atmosphere (or even that emanating from the instruments used to detect it), so it is usually observed using specially cooled mountain-top telescopes or orbiting satellite observatories. The long wavelengths of radio waves, meanwhile, present practical challenges to detection – they are usually collected using huge parabolic dishes that act similarly to reflecting telescopes.

Ultraviolet rays, conversely, have shorter wavelengths than violet light, and higher energies, while X-rays and gamma rays are still shorter and more energetic. All three of these forms of electromagnetic radiation can be harmful to living tissue and, fortunately, are mostly blocked by Earth's atmosphere. The era of high-energy astronomy has only arrived with the use of space-based telescopes, and the instruments for collecting and detecting X-rays and gamma rays bear little resemblance to the familiar telescope designs of Galileo and Newton.

The condensed idea
Telescopes reveal the hidden secrets of the Universe

03 Kingdom of the Sun

Our solar system consists of the Sun, all the objects that orbit around it, and the region of space directly under its influence. It encompasses eight major planets, five known dwarf planets, a host of moons and countless smaller objects, with both rocky and icy compositions.

For most of recorded history, the solar system consisted of just eight known objects – the Earth, Moon, Sun and five naked-eye planets: Mercury, Venus, Mars, Jupiter and Saturn. Each followed its own complex path around the sky against an apparently fixed background of more distant stars. It was not until the 16th century that Earth was widely recognized as being simply the third of six planets orbiting the Sun, and the planets' motion began to make sense (see page 6).

It was now clear that the Sun was the dominant body in our solar system, exerting a force that keeps all the planets on elliptical orbits around it. In 1687, Isaac Newton explained this as an extension of the same gravitational force that makes objects fall towards the centre of the Earth. With this model now established, astronomers were able to use geometrical techniques, enhanced in accuracy by the recently invented telescope, to measure the true scale of the solar system (see box, see page 14).

A key measurement was the average distance from Earth to the Sun, which turned out to be roughly 150 million kilometres (93 million miles). This became a convenient measurement unit in its own right, known today as

TIMELINE

1543	1610	1781
Copernicus proposes a Sun-centred view of the solar system, with Earth as one of six planets	Galileo discovers moons, previously unseen, orbiting Jupiter	William Herschel discovers a new planet beyond Saturn, later named Uranus

the astronomical unit (AU). Establishing the scale of the solar system also naturally revealed the scale of its individual planets – Venus turned out to be about the same size as Earth, Mercury and Mars significantly smaller, while Jupiter and Saturn were enormous giants in comparison.

NEW WORLDS

While 17th-century astronomers began to discover hitherto unseen moons around Jupiter and Saturn, and Saturn's magnificent ring system, the only non-planetary objects orbiting the Sun itself were thought to be comets, such as the one whose orbit was calculated by Newton's friend Edmond Halley in 1705. These were shown to be occasional visitors to the inner solar system. So in 1781, when German-born astronomer William Herschel spotted a fuzzy blue-green blob while conducting a star survey at his home in the English city of Bath, he naturally assumed it was a comet. Follow-up observations, however, revealed the truth: the object's slow movement against the stars indicated a distance of around 20 AU, suggesting that this was no comet, but a substantial planet in its own right – the world now knows it as Uranus.

> **THE SOLAR SYSTEM SHOULD BE VIEWED AS OUR BACKYARD, NOT AS SOME SEQUENCE OF DESTINATIONS THAT WE DO ONE AT A TIME.**
> Neil deGrasse Tyson

Herschel's discovery triggered a planet-hunting mania, with much interest concentrated on a perceived gap in the order of planets between the orbits of Mars and Jupiter. In 1801, this led to the discovery of Ceres (see page 40), a small world that proved not to be a full-blown planet, but the first and largest of many asteroids – rocky bodies in orbit throughout the inner solar system, but which are mainly concentrated in a broad belt between Mars and Jupiter.

While Uranus and the asteroids were found by happy accident, it was hard mathematics that led to the discovery of another major planet in 1846. In

1801	1846	1930	2016
While looking for a new planet between Mars and Jupiter, Giuseppe Piazzi discovers Ceres, the largest asteroid	Urbain Le Verrier uses irregularities in the orbit of Uranus to predict the position of an eighth planet, Neptune	Clyde Tombaugh discovers Pluto, a new world that proves to be the first known Kuiper Belt Object	Batygin and Brown claim to find evidence for a ninth major planet in the orbits of Kuiper Belt Objects

Aristarchus measures the Solar System

Third-century BCE Greek astronomer Aristarchus of Samos used an ingenious method to estimate the distances of the Moon and Sun. Realizing that the Moon's phases are caused by varying solar illumination, he measured the angle between Sun and Moon at first quarter, when exactly half of the lunar disc is illuminated, and used geometry to work out the distance to the two bodies. Thanks to measurement errors, he estimated that the Sun was 20 times more distant than the Moon (and therefore about 20 times larger). The actual figure is 400 times, but the difference was still enough to convince him that the Sun, not the Earth, must lie at the centre of the solar system.

this case, French mathematician Urbain Le Verrier carried out a close analysis of irregularities in the orbit of Uranus, pinpointing the size and location of a more distant world (now known as Neptune), which was soon spotted by German astronomer Johann Galle at Berlin Observatory.

THE HUNT FOR PLANET X

In the aftermath of Le Verrier's triumph, many astronomers became bewitched by the idea of finding new planets through mathematics. Le Verrier himself came unstuck when he predicted another planet called Vulcan, circling the Sun inside the orbit of Mercury, while others made regular predictions of a Planet X orbiting beyond Neptune. The most dedicated of these planet-hunters was wealthy amateur Percival Lowell (also an enthusiast of the so-called canals on Mars – see page 28), who set up his own observatory at Flagstaff, Arizona, and bequeathed funds for the search to continue after his death in 1916. It was at Flagstaff in 1930 that Clyde Tombaugh, a young researcher hired to carry out a new and comprehensive search for Lowell's planet, spotted a tiny dot moving against the stars on two photographic plates taken days apart. This distant world was soon named Pluto and heralded as the solar system's ninth planet.

However, Pluto's size and mass proved to be disappointingly small, and from the outset some astronomers doubted whether it should really be classed as a planet like the others. Many suspected that it was, like Ceres before, the first of an entirely new class of objects – small icy worlds orbiting beyond Neptune in what we now call the Kuiper Belt (see page 47). It was not until 1992 that the Hubble Space Telescope finally tracked down another Kuiper-Belt Object (KBO), but their numbers have since skyrocketed, with more than a thousand currently identified. Given this rate of discovery, it was inevitable that Pluto's planetary status would eventually be called into question, and in 2006 the International Astronomical Union introduced a new classification of dwarf planets that encompasses Pluto, Ceres and several other objects (see page 41).

Are there other substantial worlds still waiting to be found in the depths of the outer solar system? Current models of the solar system's birth and evolution might seem to make this unlikely (see pages 16–25), but some astronomers claim to see the orbits of certain KBOs may be influenced by unknown large planets. In 2016, Caltech astronomers Konstantin Batygin and Mike Brown made the most definitive claim so far, for a 'ninth planet' with the mass of ten Earths in a long elliptical orbit. So far, however, the only unseen objects whose existence we can be sure of are the trillions of comets of the Oort Cloud. The existence of this vast spherical halo of comets, surrounding the Sun out to a distance of about 1 light year, is revealed by the orbits of comets falling into the inner solar system.

The heliosphere

When discussing the limits of the solar system, some astronomers prefer to use not the Sun's gravitational reach, but the heliosphere, the region where the solar wind is dominant over the influence of other stars. The solar wind is a stream of electrically charged particles blowing off the surface of the Sun and extending the Sun's magnetic field across the solar system. It's responsible for phenomena such as aurorae (northern and southern lights) on various planets. The wind travels smoothly at supersonic speeds to well beyond the orbit of Pluto, but then breaks down in a region of subsonic turbulence as it encounters increasing pressure from the surrounding interstellar medium (see page 170). The outer edge of the heliosphere, where the outward flow of the solar wind comes to a halt, is known as the heliopause, and is the boundary commonly referred to when space scientists talk about missions leaving the solar system. NASA's Voyager 1 probe crossed the heliopause at about 121 AU from the Sun in August 2012.

The condensed idea
The size and complexity of our solar system keep increasing

Birth of the solar system

How did the Sun, and the motley system of planets and small bodies surrounding it, come into being? For more than two centuries, scientists have argued over various theories, but now a new idea called pebble accretion promises to finally resolve the remaining unanswered questions.

The solar system has three very distinct zones. Close to the Sun is a realm of rocky planets and asteroids dominated by 'refractory' materials with relatively high melting points, such as metals. Further out, beyond the asteroid belt, lie the giant planets and their icy moons, composed mostly of volatile chemicals that melt at lower temperatures. Most distant of all are the Kuiper Belt and Oort Cloud of small, icy bodies.

The first scientific theory of planetary origins, which sought only to explain the difference between the rocky planets and the more distant giants, was known as the nebular hypothesis. In 1755, German philosopher Immanuel Kant proposed that the Sun and planets had formed alongside each other during the collapse of a vast cloud of gas and dust. The brilliant French mathematician Pierre-Simon Laplace independently conceived a similar model in 1796. He showed how collisions within the gas cloud and conservation of angular momentum would naturally cause the planet-forming disc to flatten out and spin faster towards its centre, while forcing the resulting planets into more or less circular orbits.

TIMELINE

1734	1755	1796
Emanuel Swedenborg suggests that planets formed from collapsing gas clouds ejected by the Sun	Immanuel Kant proposes that the Sun and planets coalesced together out of an initial nebula	Laplace puts forward his own version of the nebular hypothesis, outlining the physical processes at work

A MULTITUDE OF THEORIES

By the mid-19th century, some astronomers were arguing that the spiral nebulae visible in the largest telescopes and early photographic images might be solar systems in the act of formation (see page 146). Others, however, were expressing significant doubts, in particular about the Sun's slow (c.25-day) rotation period – since our star concentrates 99.9 per cent of the solar system's mass at its very centre, surely it should spin much faster?

As these concerns took root, the nebular hypothesis was abandoned in favour of new theories. Perhaps the planets formed from a long streamer of solar atmosphere, torn off by a passing star? Perhaps they were created from captured material when the Sun did the same to another star? Or perhaps they were swept up from a cloud of 'protoplanets' in outer space?

It was not until the 1970s that astronomers began to reconsider the nebular hypothesis, thanks largely to the work of Soviet astronomer Viktor Safronov. New elements introduced to the theory allowed planets to form with much less mass in the original disc, reducing the need for a fast-spinning Sun. Key to Safronov's solar nebula disc model was the idea of collisional accretion – a process in which individual objects grow from grains of dust up to Mars-sized protoplanets, through step-by-step collisions and mergers.

COLLISIONAL ACCRETION

By the time Safronov's ideas became known outside the Soviet Union, astronomers had also learned a great deal more about the early evolution of stars themselves, and these two strands came together to build a coherent picture. As a young, hot and unstable protostar begins to shine (see page 84), it produces fierce stellar winds that blow through the surrounding

> UPON A SLIGHT CONJECTURE ... I HAVE VENTURED ON A DANGEROUS JOURNEY AND I ALREADY BEHOLD THE FOOTHILLS OF NEW LANDS. THOSE WHO HAVE THE COURAGE TO CONTINUE ... WILL SET FOOT ON THEM.
>
> Immanuel Kant

1905	1917	1978	2012
Thomas Chamberlain and Forest Moulton propose the first theory of accretion to explain how planets grow	James Jeans puts forward a tidal hypothesis to explain the origin of planets	A.J.R. Prentice shows how dust grains in the solar nebula could slow the rotation of its centre	Michiel Lambrechts and Anders Johansen propose pebble accretion as a way to rapidly form planetary cores

While the precise details of the solar system's formation are still not yet definitely known, the broad story is clear: a cloud of gas and dust began to collapse under its own gravity [1], flattening out into a disc with a bulging centre [2]. The Sun formed at the centre, with the solid cores of protoplanets in orbit around it [3]. These mopped up material from their surroundings to produce today's major planets [4].

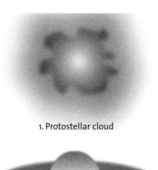

1. Protostellar cloud

3. Formation of planetary cores

2. Protoplanetary disc

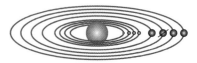

4. Planets clear their orbits of material

nebula, alongside strong radiation that raises the temperature of the nebula's inner regions. This has the effect of causing volatile icy material close to the star to evaporate, and then blow outwards leaving the dusty refractory material behind. Random collisions over a few million years see these particles grow from dust grains to pebbles to small asteroids. Once they are large enough to exert moderate gravity, the process snowballs in an effect known as runaway accretion. The growing bodies, known as planetesimals, pull more and more material towards themselves, clearing most of the surrounding space until a few dozen worlds, perhaps the size of our Moon, remain. Collisions between these protoplanets give rise to a smaller number of rocky planets, while the heat unleashed by impacts causes them to melt, allowing their interiors to differentiate and their crusts to settle into a spherical shape.

Further out in the star system it is colder. Volatile ices remain frozen and gas tends to linger, leaving much more planet-building material. The process of planet formation proceeds in more or less the same way on a much grander scale, resulting in planets with larger solid cores, which then pull in surrounding gas to form deep, hydrogen-rich atmospheres. On the outer edges of the planet-forming zone, material is too thinly spread to form large planets, resulting in a proto-Kuiper Belt of icy dwarf worlds.

Safronov's theory has held sway for more than four decades. Backed up by the discovery of planet-forming discs around many other stars, it is widely agreed to be accurate when it comes to the bigger picture. However, some astronomers have recently begun to suspect that it's not the whole story. In particular, there are doubts about Safronov's two-body collision model, and there's growing evidence that many worlds in the solar system did not undergo the kind of complete melting required by Safronov's repeated planetesimal collisions. Just as importantly, scientists have realized there's a gap in the chain of growth. On a small scale, tiny static electric charges on dust grains would draw them together, while mutual gravitational attraction would attract large-scale objects together. But how do boulder-sized objects stick together as they grow from one stage to the other? The solution to these problems may lie in a remarkable new theory called pebble accretion (see box), which involves large numbers of small objects coalescing at the same time.

Pebble accretion

Recently, experts on planet formation have honed a new theory to explain several outstanding mysteries in planet formation: not only how accreting bodies crossed the size threshold from small to large scale, but also how the gas giants grew their cores quickly enough to retain rapidly disappearing gas, and why the terrestrial planets seem to have formed at remarkably different times. Pebble accretion suggests that the early solar system rapidly evolved huge drifts of small solid fragments, slowed down and corralled by their motion through surrounding gas. Within just a couple of million years of the Sun's formation, these drifts grew large enough to become gravitationally unstable, collapsing to form Pluto-sized planetesimals in a matter of months or years. The gravity of these worlds then rapidly drew in the remaining pebbles from their surroundings, leaving perhaps a couple of dozen Mars-sized worlds. The giant planets were therefore able to start accumulating their envelopes of gas and ice early on, while Mars was fully grown. Only the larger terrestrial planets, Earth and Venus, required a final phase of Safronov-style collisions, over the next hundred million years or so, to reach their present size.

The condensed idea
Planets grow through small objects merging together

Planetary migration

Until recently, most astronomers believed that the planets of our solar system had followed stable orbits throughout their history. But new advances in computer modelling suggest that the early days of the solar system involved a vast game of planetary pinball whose aftermath can still be seen today.

B efore the discovery of the first exoplanets in the mid-1990s (see page 96), astronomers tended to believe that alien solar systems would be more or less like our own, with planets following near-circular, stable orbits around their stars. However, just as the past two decades of research have shown that planetary systems are a lot more varied than previously thought, so advances in simulation and modelling based on the collisional accretion model of our solar system's formation have suggested that planet-forming material would have petered out around the orbit of Saturn. So where did Uranus and Neptune come from? In an effort to answer these and other questions, in 2005 a group of astronomers put forward a remarkable new theory that the first few hundred million years of our solar system's lifetime saw radical shifts in the distribution of the planets.

WORLDS ON THE MOVE

Theories of planets shifting orbits have gone through periods of popularity since the 19th century, while being roundly dismissed as pseudoscientific nonsense by the astronomical establishment. Indeed, the ideas of 'independent scholars' such as Immanuel Velikovsky, who posited planets ricocheting around the solar system in relatively recent times as an

TIMELINE

1950	1974	2005
Immanuel Velikovsky's *Worlds in Collision* attempts to explain historical events through a pseudoscientific planetary migration theory	Tera, Papanastassiou and Wasserburg discover evidence for the Late Heavy Bombardment in lunar rock samples from the Apollo missions	The Nice Model is launched through the publication of three scientific papers in *Nature*

explanation for many mythological and historic events, are easily dismissed. But the so-called Nice Model, named after the French city where many of it developers worked at the Observatoire de la Côte d'Azur, is a very different prospect. The Nice Model is a set of interlocking proposals based on computer modelling the evolution of the early solar system in order to resolve some longstanding mysteries. In little more than a decade, it has opened up a new and exciting field of research into the neglected field of solar system dynamics.

THE NICE MODEL

The model postulates that shortly after its formation, the outer solar system was very different from its current state. All four giant plants were more tightly packed, with near-circular orbits inside the current orbit of Uranus (about 20 AU from the Sun). Furthermore, Neptune, now the outermost planet, orbited closer to the Sun than Uranus. Extending beyond the major planets lay a proto-Kuiper Belt – a disc of icy objects whose largest worlds were about the size of today's dwarf planets, and which were contained within the current orbit of Neptune.

Computer simulations indicate that such an arrangement of giant planets would have been stable for about 500 million years, before a series of close encounters between Uranus and Neptune disrupted their orbits and pulled them into elongated elliptical paths. These eccentric orbits soon brought them closer to the much larger planets Jupiter and Saturn, whose powerful gravity flung them out onto much larger, though still elliptical, paths around the Sun, and threw Neptune beyond Uranus for the first time. It was probably during this event, too, that Uranus acquired its current remarkable tilted axis, which has the gas giant planet rotating on its side, rather like a rolling ball in contrast to the 'spinning top' motion of the other planets.

> **IT WAS A VERY VIOLENT, SHORT-LIVED EVENT LASTING JUST A FEW TENS OF MILLION OF YEARS.**
> Hal Levison

2011	2011	2016
David Nesvorny proposes a fifth giant planet in the early solar system as a means of resolving problems with the Nice Model	Some of the original Nice Model researchers propose the Grand Tack of Jupiter to explain the small size of Mars	Planet-hunter Mike Brown claims to have found evidence for the exiled fifth giant planet in the orbits of Kuiper Belt Objects

The Late Heavy Bombardment

Thanks to the radiometric dating of lunar rocks brought back by the Apollo astronauts, many astronomers believe that the inner solar system went through a traumatic phase about 3.9 billion years ago, in which worlds such as the Moon suffered an intense bombardment of large planetesimals. On the Moon, craters left behind by these impacts were later filled with lava from volcanic eruptions, creating the smooth, dark lunar 'seas' that dominate the lunar near side today.

First recognized in the late 1970s, this Late Heavy Bombardment was for a long time assumed to be simply a mopping-up phase at the end of planetary accretion, but more recent evidence suggests that the main phase of planet formation came to an end much earlier. Instead, perturbations created as the giant planets shift their orbits in the Nice Model are now the favoured explanation. However, some sceptics have suggested that the bombardment never happened on the scale that some envisage, arguing instead that all of the impact melt samples collected by Apollo astronauts actually originated from a single vast impact event

DISLODGING THE KUIPER BELT

The new orbits of Uranus and Neptune, however, sent them ploughing straight through the proto-Kuiper Belt, where further encounters with small, icy worlds helped to circularize the ice giants' orbits at greater distances from the Sun. Many of the smaller worlds were ejected further out into a region known as the scattered disc, while others were sent plunging towards the inner solar system, where they caused the cataclysmic event known as the Late Heavy Bombardment (see box, left).

The Nice Model is intriguing, not least because it promises to solve mysteries such as the tilt of Uranus, the location of the gas giants and the Late Heavy Bombardment. It can also provide mechanisms for capturing the Trojan asteroids that share an orbit with Jupiter, Uranus and Neptune. But the model is not perfect: it has difficulty in explaining how Jupiter ended up with its current large family of captured moons, and the combined gravitational influence of Jupiter and Saturn as they pass through a period of orbital resonance (with frequent close encounters) could also have caused problems. In fact, some simulations show violent effects such as the complete ejection of Mars and destabilization of the other planets – issues that are big enough for the model to be tweaked substantially. Similarly, the frequency with which modelled encounters between Jupiter and Uranus or Neptune end with the smaller world being kicked out of the solar system completely has led some astronomers to argue for an early solar system with *three* ice giants.

Despite these problems, the Nice Model or something like it remains a key part of current ideas about our solar system's history. And other astronomers are applying similar thinking to other questions. For example, why did Mars never grow into a body the size of the Earth, and where did our own planet's abundant water come from? The answer to both of these questions may lie in the Grand Tack, a hypothetical path taken by a newly formed Jupiter in the gas-rich environment of the very early solar nebula (see page 19). According to this theory, interaction with the nebula caused Jupiter's orbit to drift first inwards, and then outwards. In the process, the giant planet's gravity would have disrupted (and stolen) much of the planet-forming material around the orbit of Mars, and later enriched the outer asteroid belt with icy bodies from further out in the solar system. Once dislodged, these could have rained down upon Earth, bringing with them the water that makes our planet habitable today.

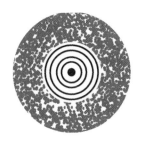

4.5 billion years ago, the giant planets were constrained within Saturn's present-day orbit, surrounded by a large proto-Kuiper Belt.

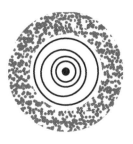

4.1 billion years ago, the influence of Jupiter and Saturn kicks Neptune and Uranus into elliptical orbits that begin to disrupt the proto-Kuiper Belt.

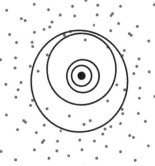

4.1–3.8 billion years ago, Neptune and Uranus reach maximum eccentricity and switch their order from the Sun. Kuiper Belt Objects are flung in all directions, bombarding the inner solar system.

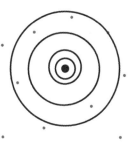

By 3.5 billion years ago, the orbits of Uranus and Neptune have become more or less circular and the solar system has taken on its present configuration.

The condensed idea
The planets have not always followed the same orbits

06 Birth of the Moon

Compared with the majority of satellites in our solar system, Earth's Moon is very different. Its huge size compared to our own planet suggests it must have had a very unusual origin. But the truth about that origin has only become clear since the 1980s, and some questions still remain unanswered.

E arth's Moon is huge – with about a quarter of the Earth's diameter, it's by far the largest satellite of a major planet compared to its parent planet. But its strange nature became clear only gradually in the centuries after the invention of the telescope. Theories of the origin of the solar system (see page 16) could neatly explain the moon families of giant planets as leftover debris that coalesced in orbit (a smaller-scale version of the birth of the solar system itself), but it soon became clear that this 'co-accretion' model failed when it came to the Earth. Aside from the basic question of why Earth alone had enough excess material to make a massive satellite, it was clear that the angular momentum of the combined Earth–Moon system is very high compared to the other terrestrial planets, something that would not be expected if the Moon had formed from a slowly spinning disc.

EARLY THEORIES

In the quest for answers, 19th-century astronomers came up with two theories – capture and fission. The capture model suggests that the Moon formed somewhere else in the solar system and was subsequently captured by Earth during a close encounter. It failed to explain why the Moon's density is significantly lower than Earth's, however, and required a

TIMELINE

1946	1969–72	1974
R.A. Daly first suggests a giant impact origin for the Moon	Manned Apollo landings return 382 kg (842 lb) of Moon rock for analysis on Earth	Hartmann and Davis model possible origins for an impacting body

highly improbable encounter scenario. It's a lot tougher for a small planet to capture a large satellite than it is for a giant planet to do so (and even then, we only know of one such large, captured satellite in the outer solar system – Neptune's icy moon Triton).

The fission hypothesis, meanwhile, was first promoted by English astronomer George Darwin (son of Charles). Darwin studied the tidal forces between the Earth and Moon and showed that our satellite's orbit is spiralling slowly outwards by about 4 cm (1.6 in) per year, while Earth's rotation is gradually slowing down. He correctly concluded that the Earth and Moon were once much closer together, and argued that they originated as a single, rapidly rotating body: the material that formed the Moon was flung off the bulging equator of the primordial Earth, before coalescing in orbit. Darwin even claimed that the Pacific Ocean basin marked the still-visible scar of this violent separation. The theory enjoyed several decades of popularity in the early 20th century, before further studies of the forces involved concluded around 1930 that the scenario was essentially impossible.

In the 1970s, new evidence finally arrived in the form of rock samples brought back by the Apollo Moon missions. These showed that lunar rocks were extremely dry – not only is water absent from the upper layers of the crust, but so are the hydrated minerals found on Earth. The rocks were also severely depleted in terms of volatile (low-melting-point) elements such as potassium, lead and rubidium, compared to both Earth and models of the local primordial solar nebula. Conversely, the Moon proved richer in iron oxide than the Earth's own mantle, despite having only a small iron core.

THE BIG SPLASH
These findings inspired renewed interest in a neglected theory put forward by Canadian geologist Reginald Aldworth Daly as early as 1946: the giant

1976	1994	2012
Cameron and Ward model the dynamics of a Moon-forming impact	NASA's Clementine mission reveals unexpected survival of volatile elements in lunar crust	Evidence for extreme similarity of Earth and Moon materials inspires new origin theories

Modelling Theia

As evidence has mounted that the Moon formed from essentially Earth-like materials, the origins of the impactor planet Theia have become ever more constrained. Because the ratio of elemental isotopes across the solar nebula was so sensitive to their distance from the Sun (see page 18), it's clear Theia must have formed very nearby, from essentially the same mix of materials. However, Earth's own gravity would have disrupted the formation of any objects in nearby space, so where could Theia have come from? One theory is that Theia formed at either the L4 or L5 Lagrangian points – gravitational sweet spots on the same orbit as Earth, but 60 degrees ahead or behind the larger planet, where the influence of Earth was minimized. Here, Theia could have grown to about 10 per cent of Earth's mass before its stable orbit was eventually disrupted and it fell on a path to inevitable collision. This could account for the similarity between the raw materials in both worlds (though some doubt that even this explanation is enough). What's more, since the two worlds were moving in very similar orbits, the energy of the collision would have been much lower, perhaps explaining the survival of remnant volatile elements in the Moon today.

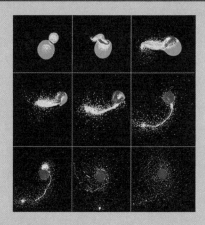

A computer simulation models the formation of a proto-Moon in mere hours after the collision between a Mars-sized planet and a young Earth with roughly 90 per cent of its present-day mass.

impact hypothesis. In this modified version of the fission theory, much of the material to form the Moon came from the Earth, ejected not by rapid rotation, but from an interplanetary collision with a planet-sized body.

William K. Hartmann and Donald R. Davis of the Planetary Science Institute in Tucson, Arizona, showed the plausibility of other bodies forming close to the primordial Earth, while A.G.W. Cameron and William Ward at Harvard modelled the impact itself, showing that it probably involved a Mars-sized body striking the Earth at a tangent. Such an event would result in large amounts of the impactor, along with a substantial chunk of Earth's mantle, being melted and ejected into orbit, but would have seen most of the impactor's iron core absorbed by the Earth. The intense heat of the impact would explain the lack of water and other volatiles in lunar rock.

OUTSTANDING QUESTIONS

Planetary scientists have widely accepted the giant impact hypothesis since the 1980s. The collision is thought to have occurred around 4.45 billion years ago, with the Moon coalescing rapidly within hours of the collision. The impacting planet has even been given an unofficial name, Theia, after the mother of the Moon goddess Selene in Greek mythology. However, there are still significant unanswered questions. Closer study of lunar rock samples has shown that they are not quite as depleted of volatiles as they should be after such a violent collision. In fact, temperatures do not seem to have risen above about 950°C (1740°F). Meanwhile, the mix of isotopes (atoms of the same element but with different weights, whose relative abundances are an extremely sensitive indicator of the provenance of raw materials in the solar nebula) have shown an extremely close match to the Earth – so close as to suggest no contribution from Theia.

> **THEIA WAS THOROUGHLY MIXED INTO BOTH THE EARTH AND THE MOON, AND EVENLY DISPERSED BETWEEN THEM.**
> Edward Young

In order to address these problems, various theories have been advanced, perhaps the most radical of which involves Earth and Moon coalescing *together* out of an initial collision between two much larger bodies, each about five times the size of Mars. In 2016, meanwhile, a team led by Edward Young of the University of California at Los Angeles put forward new evidence from chemical comparisons that Theia and the Earth collided in a head-on collision that mixed their materials very thoroughly. It seems clear that the Moon's origin – and perhaps Earth's own – was messier and more complex than the simple giant impact hypothesis would suggest.

The condensed idea
Our satellite was born out of an interplanetary collision

07 **Water on Mars**

A series of discoveries has transformed our understanding of the famous Red Planet Mars. Long seen as a cold, arid desert, it now seems clear that water lies just below the surface not only as abundant ice, but also in liquid form. What's more, it's possible that water occasionally becomes far more widespread.

A stronomers have been fascinated by the possibility of water flowing on the surface of Mars ever since Italian Giovanni Schiaparelli reported seeing narrow channels (*canali*) linking the darker areas of its surface in 1877. Misunderstood to be artificial canals in the English-speaking world, similar channels were reported by many other observers, and sparked a wave of speculation into the possibility of intelligent life on Mars. Even when improved observations and experiments in the early 1900s showed that the canals were nothing more than an optical illusion, the idea of Mars as a warm world with a fairly dense atmosphere and surface water persisted through much of the 20th century. It was only in the mid-1960s, when NASA's Mariner space probes flew past the planet, that the truth was revealed: Mars' sparse atmosphere seemed to have left it as nothing more than a cold Moon-like world of craters and endless red dust.

A WET PAST – AND PRESENT?
From that late 1960s low point, however, successive missions to the Red Planet have seen it recover some of its former glamour, revealing ever-more Earth-like facets. Mariner 9, which arrived at Mars in November 1971, began the trend – although it had to wait two months for a huge planet-wide

TIMELINE

1877	1965	1972
Giovanni Schiaparelli mistakenly reports the existence of water channels on Mars	Mariner 4 becomes the first spacecraft to fly past Mars, sending back images that suggest an arid, dead world	Mariner 9 discovers evidence of ancient floods and dried-up riverbeds on the Martian surface

dust storm to clear before it could begin its first survey from orbit. While most famous for identifying the enormous Martian volcanoes and the vast canyon-like fault of the Valles Marineris, it also revealed large parts of the Martian surface that were far less heavily cratered and showed signs of possible water action in the distant past. These included sinuous valleys resembling river valleys on Earth and flattened scablands, apparently created during catastrophic floods.

Mars, it seemed, might have been wet once upon a time, but what about now? While the Viking orbiters of the mid-1970s bolstered the evidence for a watery Martian past, and their companion landers found signs that surface rocks had been exposed to moisture or even submerged in the distant past, there was little evidence of water surviving to the present, except perhaps deep-frozen in the Martian ice caps.

> **THERE IS LIQUID WATER TODAY ON THE SURFACE OF MARS.**
> Michael Mayer, NASA, 2015

That picture began to change rapidly from the late 1990s. Mars Global Surveyor (MGS), an orbiting satellite capable of photographing the planet in far more detail than the Vikings, detected signs of buried ice at lower latitudes away from the polar caps, and in 2002, NASA's Mars Odyssey probe revealed evidence for huge deposits of frozen ice in the Martian soil. So much, in fact, that at latitudes above 55° in each hemisphere, 1 kilogram (2.2 lb) of soil is thought to contain about 500 grams (18 oz) of water. In 2008, NASA's Phoenix probe landed close to the north polar cap and confirmed the presence of ice in the soil.

But does water flow on the Martian surface today? At face value, it seems unlikely – while Martian surface temperatures can reach up to 20°C (68°F), they mostly stay well below freezing, while the thin carbon dioxide atmosphere, exerting just one per cent of Earth atmospheric pressure, would cause any exposed water to rapidly boil away.

2002	2006	2015
The Mars Odyssey mission discovers huge amounts of water ice in the soil across much of the northern hemisphere	Mars Reconnaissance Orbiter (MRO) finds fresh gully features that may be formed by water-related processes	MRO finds fresh hydrated minerals in recurring slope lineae, confirming liquid water close to the surface

Martian Milanković cycles

In the early 1920s, Serbian scientist Milutin Milanković made a remarkable proposal to help explain long-term climate cycles during Earth's most recent ice age. He suggested that the amount of sunlight heating our planet is altered slowly but significantly as several of Earth's orbital parameters vary over time, due to the influence of other planets. Now planetary scientists are starting to wonder if similar Milanković cycles could be responsible for long-term changes to the Martian climate. The specific changes in question are:

• A 124,000-year variation in the planet's tilt between angles of 15° and 35°, affecting the severity of the seasons.
• A 175,000-year wobble or 'precession' in the direction of the planet's tilted axis, affecting how much each hemisphere is affected by seasonal changes.
• 100,000- and 2.2-million year cycles in the eccentricity of the Martian orbit, which ranges from an almost perfect circle to a marked ellipse, tending to exaggerate or muffle the effect of seasons.

Astronomers studying the annual ice layers of the Martian polar caps think they may have found signs of variation that could offer a good match to some of these cycles.

MYSTERIOUS GULLIES

One landmark discovery from Mars Global Surveyor opened up the debate once again – photographs of fresh, gully-like features on the slopes of a valley called Gorgonum Chaos in the planet's mid-southern latitudes. Apparently originating from a layer just below the surface, many wondered if they were signs of liquid water seeping out from a buried aquifer, carving a channel through the surface dust before evaporating. Similar features on Earth are the work of flowing water, but more cautious voices put forward other potential causes, such as explosive evaporation of carbon dioxide ice exposed in an underground layer.

The mystery deepened in 2006, when NASA's newly arrived Mars Reconnaissance Orbiter (MRO) discovered gullies in areas that had no such features in the MGS images taken a few years earlier. Clearly gully formation is an active and ongoing process. The new gullies formed at similar latitudes to Gorgonum Chaos, and mostly on steep south-facing slopes. One theory is that snow tends to accumulate in such areas (which receive little sunlight in winter) and the gullies are the result of a spring thaw. Definitive evidence that the gullies are formed by water has remained elusive, but focus on the mid-latitudes where they are concentrated has yielded more conclusive results. In 2011, NASA announced the discovery of recurring slope lineae (RSLs) in many of the same locales – long dark streaks that extend down slopes such as crater walls through the Martian summer, and disappear in winter. In contrast to the gullies, RSLs concentrate on equator-facing slopes that receive the most sunlight throughout the year and are therefore relatively warm, at temperatures around −23°C (−9°F).

Although initially thought to be caused by briny water (water with a significant amount of salt, which lowers its melting point), the RSLs are not simply damp areas of the soil. Instead, they seem to be rough patches that somehow smooth themselves out and disappear during the cold season. Clinching proof that water is indeed the cause was found in 2015, when instruments aboard MRO confirmed that the spread of the lineae is accompanied by the formation of hydrated mineral salts. The new consensus is that the lineae are created by briny water flowing just beneath the surface and disturbing loose overlying soils. Mars, it seems, is not the dry desert we previously thought, and this raises intriguing questions about the prospects for Martian life (see page 48).

Global warming?

Recent evidence from space probes suggests that Mars could be moving from cold and dry conditions to warmer and wetter ones under our very noses. Comparisons between global average temperatures measured by the Viking orbiters of the 1970s and those recorded in the mid-2000s show a 0.5°C (0.9°F) rise over three decades, coinciding with shrinking ice at the polar caps (pictured below). One important factor in this apparent global warming may be the release of huge plumes of methane, discovered in 2009 above the warmest areas of the planet and thought to be escaping from melting subterranean ice. Although short-lived in the atmosphere, methane is a powerful greenhouse gas and could act to speed up the rate of warming.

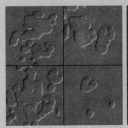
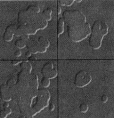

1999   2001

The condensed idea
Mars might be a desert, but it's not bone dry

08 Gas and ice giants

Astronomers have only recently discovered that there are two kinds of giant planet in the outer solar system – enormous, low-density gas giants Jupiter and Saturn, and the smaller, denser ice giants Uranus and Neptune. But how exactly did these worlds form, and why are the two types so different?

Until the 1990s, the terms 'gas giant' and 'giant planet' were synonymous. The largest worlds of the solar system were all assumed to have a similar structure, with a solid core (perhaps about the size of Earth) surrounded by a deep atmosphere mostly composed of the lightweight elements hydrogen and helium. The distinctive colours seen in the planets' upper atmospheres were linked to relatively small quantities of other chemical compounds, while 1,000 kilometres (620 miles) or more below the visible surface, the gaseous elements were transformed by rising pressure into an ocean of liquid hydrogen.

DISCOVERING THE ICE GIANTS

The picture began to change as researchers analysed data from the Voyager 2 space probe's flybys of Uranus and Neptune (in 1986 and 1989, respectively). One key piece of evidence for a major internal difference came from the outer planets' magnetic fields. They were relatively weak, tilted sharply in relation to the planets' axes of rotation and offset from the centre of each planet. In stark contrast, the fields around Jupiter and Saturn are far more powerful, centred within each planet and closely aligned with their rotational poles.

TIMELINE

1665	1690	1781	1846
Giovanni Domenico Cassini makes the first observation of Jupiter's Great Red Spot	Cassini measures varying rotation of Jupiter's features, revealing that it is not a solid body	William Herschel discovers Uranus, the first new planet in the solar system	Johann Galle discovers Neptune, following a prediction by Urbain Le Verrier

Jupiter's and Saturn's magnetism could already be explained through a dynamo effect, produced by a churning layer of liquid metallic hydrogen around each planet's solid core. Under extreme temperature and pressure, molecules in the liquefied gas split apart to create an ocean of electrically charged ions. The fact that this was not happening at Uranus and Neptune suggested that liquid hydrogen was simply not present in large quantities at great depths.

> IT IS NOT AMAZING THAT CHEMISTRY LIKE THIS HAPPENS INSIDE PLANETS, IT'S JUST THAT MOST PEOPLE HAVEN'T DEALT WITH THE CHEMICAL REACTIONS THAT CAN OCCUR.
>
> Laura Robin Benedetti

Instead, scientists soon concluded, the interiors of the outer giants were probably dominated, like so much of the outer solar system, by water and other volatile chemical ices. Hydrogen-rich outer layers give way, a few thousand kilometres down, to a mantle of relatively heavy compounds – predominantly water, ammonia and methane. So while hydrogen and helium account for more than 90 per cent of the mass of Jupiter and Saturn, they contribute only 20 per cent of the mass of Uranus and Neptune.

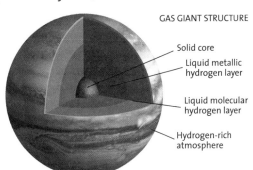

GAS GIANT STRUCTURE

Solid core

Liquid metallic hydrogen layer

Liquid molecular hydrogen layer

Hydrogen-rich atmosphere

Despite their name, however, it would be a mistake to think of ice giants as deep-frozen balls of solid matter. In this case, ice is just shorthand for a mix of volatile compounds – water, methane and ammonia together forming a churning liquid ocean beneath the outer hydrogen atmosphere. Weak electric currents in this mantle zone are thought to be responsible for the planets' curious magnetism.

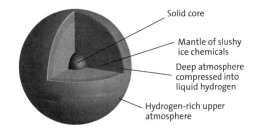

ICE GIANT STRUCTURE

Solid core

Mantle of slushy ice chemicals

Deep atmosphere compressed into liquid hydrogen

Hydrogen-rich upper atmosphere

1952	1972	1986–89	2014
Science-fiction author James Blish coins the term 'gas giant'	NASA launches the first Pioneer space probes to Jupiter and Saturn	Voyager 2 flies past Uranus and Neptune, finding evidence that they have a more icy composition than the inner giants	Lambrechts, Johansen and Morbidelli put forward a pebble accretion model to explain the formation of the giant planets

GIANT ORIGINS

The question of how these strange intermediate worlds formed, however, is a puzzle for planetary scientists. The traditional accretionary model of planet formation (see page 16) has difficulties explaining anything forming so far out in the solar system (Uranus orbits at around 19 AU from the Sun, Neptune at around 30 AU). One problem is that planetesimals (the medium-sized, intermediate step in planet formation) orbiting this far from the Sun would only need a small gravitational kick to be ejected from the solar system completely. In fact, thanks to the gravity of Jupiter and Saturn orbiting closer in, they were far more likely to get that kick than they were to collide and accrete in large numbers.

One possible solution relies on so-called disc instability, a model in which the giant planets do not grow through accretion, but instead collapse very suddenly out of large-scale eddies in the solar nebula. In this way, proponents argue, a planet could coalesce in as little as a thousand years. The alternative is that the giants all formed in less perilous conditions closer to the Sun, but Uranus and Neptune later went through a period of orbital change that saw them shunted into their current orbits (the basis of the Nice Model of planetary migration – see page 20). In this situation, new models show their cores could have formed quite rapidly through pebble accretion (see page 19), allowing them to gain enough gravity to draw in gas from their surroundings in the ten million years or so before it was blown away by radiation from the strengthening Sun.

Neither model, however, offers a good explanation for why the gas and ice giants should be so different. Various mechanisms have been suggested: for example, the disc instability model posits that the giant planets all began life even bigger, before losing most of their atmospheric envelopes under assault from the fierce ultraviolet radiation of other nearby stars (a process called photoevaporation that has been seen around today's newborn stars – see page 84). The more massive Jupiter and Saturn were better able to withstand this trial and therefore retained more hydrogen, while Uranus and Neptune had most of it stripped away.

Recent work on pebble accretion provides another possibility, in which a small initial variation between growing planetary cores creates a big difference among the eventual planets. This 'threshold mass' model suggests that the rapid growth of planetary cores from centimetre-scale pebbles generates heat that prevents gas from collapsing onto the core. If it reaches a certain mass, however, the core's gravity carves out a gap in the disc of orbiting pebbles, cutting off its own food supply. As the core now begins to cool, it rapidly accumulates gas from its surroundings, growing into a gas giant. Ice giants, meanwhile, are planets whose cores, forming just slightly further out in the nebula, never reached the threshold, or reached it too late to cling onto much of the infant solar system's fast-disappearing hydrogen. They, therefore, retain a composition that owes much to the original icy pebbles of the outer solar nebula.

Diamond rains?

One of the most eyecatching theories to have emerged from recent studies of gas- and ice-giant structure is that both types of planet might create rains of crystalline carbon (diamond) deep within their interiors. In 1999, a team of researchers from the University of California at Berkeley compressed liquid methane, found in large amounts within both Uranus and Neptune, to more than 100,000 times Earth atmospheric pressure, while simultaneously heating it to around 2,500°C (4,530°F). The result was a dust of microscopic diamond particles suspended in a mix of oily hydrocarbon chemicals. Because conditions inside the ice giants do not get hot enough to melt diamond, any particles produced would slowly sift down through the planet's liquid inner layers to settle on its solid core.

In 2013, scientists from NASA's Jet Propulsion Laboratory calculated that something even more impressive might happen in the atmospheres of the larger gas giants. Here, powerful lightning bolts can cause methane to disintegrate into carbon soot high in the atmosphere, and as the soot gradually tumbles down it would be compressed to form fingertip-sized diamonds. Unlike on ice giants, however, these diamonds would not survive their descent through the planet. Below depths of about 30,000 kilometres (19,000 miles), temperatures becomes so extreme that they would melt, perhaps even forming a layer of liquid carbon with icy 'diamond bergs' floating on it.

The condensed idea
Gas and ice giant planets have very different compositions

Ocean moons

Each of the giant plants in the outer solar system is orbited by a large family of icy satellites, many of them formed at the same time and from the same materials as the planets themselves. But there's growing evidence that several of these moons are not as deep-frozen as they appear to be at first glance.

The largest moons of the outer solar system were discovered soon after the invention of the telescope in the early 17th century – the four large satellites around Jupiter in 1610 and Saturn's giant moon Titan in 1655. Many other moons have since been found around both of these planets, and Uranus and Neptune also proved to have satellite families of their own. But the nature of these moons remained unknown until the mid-20th century, when spectroscopy (see page 60) using advanced ground-based telescopes found evidence for large amounts of water ice on many of their surfaces. As a general rule, the rock content of a moon dwindles with distance from the Sun, but ice is a major component of almost every major satellite. This is only to be expected, since these worlds all were born beyond the snow line of the early solar system, in a region where ice dominated the raw materials of planet formation.

EARLY THEORIES

In 1971, a couple of years before the first space probes reached Jupiter, US planetary scientist John S. Lewis published the first detailed analysis of what they might expect to find among the Jovian moons. He argued that the slow decay of materials such as radioactive uranium within the rocky

TIMELINE

1971	1979	1979
Lewis argues that some moons may be heated enough by radioactive decay to maintain liquid oceans under an icy crust	Peale proposes tidal heating as a mechanism may drive geological activity on Jupiter's larger moons	Voyager 1 discovers volcanic activity at Io, and an icy crust on Europa

component of these moons could generate significant amounts of heat – perhaps enough to melt the icy material around a rocky core and create a global ocean capped by a frozen crust. The idea began to gain popularity when images from Pioneers 10 and 11 confirmed that Jupiter's three outer large moons – Europa, Ganymede and Callisto – shared a general icy appearance (although there were notable differences). However, the innermost moon, Io, looked starkly different with no signs of water in its composition. Io presented an obvious problem and various explanations for its striking difference were put forward throughout the 1970s.

Then in 1979, mere days before the Voyager 1 probe's flyby of Jupiter – which was scheduled to include much closer flybys of the Jovian moons – a daring new explanation for the differences emerged. Stanton J. Peale of the University of California at Santa Barbara argued, along with two NASA-based colleagues, that Jupiter's powerful gravity exerts a tidal heating effect on its innermost satellites. Even though their orbits are nearly circular, slight differences in distance cause the shape of the inner moons (notably Io and Europa) to flex with each orbit. This generates friction within their rocks, and heats them far more than radioactive decay could on its own.

> **ALL THESE WORLDS ARE YOURS, EXCEPT EUROPA. ATTEMPT NO LANDING THERE.**
> Arthur C. Clarke,
> *2010: Odyssey Two*

Most importantly, Peale suggested that Io should show signs of volcanic activity on its surface, a prediction that was borne out when Voyager 1 sent back photos of lava flows and a huge plume of molten sulphur compounds erupting into space above the moon. Any water Io might have once contained, it seemed clear, had evaporated long ago.

WATER AT JUPITER
The discovery of powerful tidal heating revolutionized ideas about environments in the outer solar system, with significant implications for

1995–2003	2005	2013
Galileo probe measurements reveal liquid water layers on Europa, Ganymede and Callisto	The Cassini probe discovers a vast plume of water ice emerging from Enceladus	The Hubble Space Telescope detects water vapour over Europa's south pole

Europa. Voyager images confirmed the presence of a thick icy crust, but also showed that the surface was clearly being renewed and rearranged on a short timescale (in geological terms). Stained with impurities apparently bubbling up from beneath, Europa's crust looked more like compressed pack ice than a smooth glacial shell. The best explanation for these features is that volcanic eruptions beneath the crust release heat, creating a global ocean of liquid water upon which the solid crust slowly shifts and churns.

One of the key features of the tidal heating model, however, is that its effects drop rapidly with distance from the parent planet, so it seemed unlikely to affect the more distant Ganymede or Callisto. Indeed, Voyager images suggested Ganymede might have gone through a Europa-like phase in its early history before freezing solid, while Callisto's interior had probably never melted at all. It was something of a surprise, therefore, when the Galileo mission to Jupiter found magnetic evidence for subsurface oceans on both moons (see box, left).

Even more astonishing discoveries awaited the Cassini spacecraft when it entered orbit around Saturn in 2004. One of the mission's prime targets was Saturn's giant moon Titan, a frozen world on which methane seems to play a similar role to water on Earth. Even so, the moon may hide a buried mantle of liquid water and ammonia deep beneath its surface (see box, opposite).

THE PLUMES OF ENCELADUS

The unexpected highlight of Cassini's mission, however, turned out to be the

Magnetic evidence

Apart from looking for surface activity or traces of water from geological history, planetary scientists can search directly for subterranean oceans by studying the magnetic fields of various moons. If a satellite has a layer of electrically conductive, mobile material beneath its surface, then as it moves through the magnetic field of its parent planet, motions called eddy currents are generated in the conducting layer. These in turn create a distinct induced magnetic field around the moon, which can be detected by magnetometers carried on passing space probes. The induced field is quite distinct from any intrinsic magnetic field such as that due to an iron core, and its shape and strength can reveal the depth and electrical properties of the conducting layer. Induced fields have been discovered not only around Europa and Enceladus, but also around Jupiter's largest satellites Ganymede and Callisto, and Saturn's giant moon Titan, all indicating salty, highly conductive oceans at varying depths.

much smaller moon Enceladus. With a diameter of just 504 kilometres (313 miles), this satellite has one of the brightest surfaces in the solar system, and a handful of photos from the Voyager probes gave the appearance of a landscape blanketed in fresh snow. It was still a surprise, though, when during an early flyby Cassini flew straight through a vast plume of water ice crystals erupting close to the moon's south pole. Some of the plume's contents escape into space to form a faint outer ring around Saturn, but most fall back onto Enceladus itself.

More than 100 individual plumes have now been identified, mostly erupting along ridgelike features known as tiger stripes. These are weak areas of the crust where cracks allow the salty liquid water below to boil away into space. Tidal heating once again seems to be the cause; in this case, heat is generated because Enceladus' orbit is prevented from becoming perfectly circular by the pull of the next moon out, Dione. In contrast to Europa, conditions on this moon seem to permit water remarkably close to the surface, making Enceladus one of the most promising places in our solar system to look for life.

Cryovolcanism

Enceladus and Europa may be the only worlds with sufficient tidal heating to melt pure water, but many of the other ocean moons in the outer solar system may owe their liquid environments to the presence of other chemicals. It's a well known fact that salt in the Earth's oceans lowers the freezing point to around −2°C (28°F), and there is good evidence that many extraterrestrial subterranean oceans are just as salty as those on Earth. However, the presence of ammonia has an even more dramatic effect, lowering the freezing point by tens of degrees – enough for water to remain liquid even with weaker tidal heating, and to vaporize in the kind of geyser-like plumes seen on Enceladus. What's more, because an ammonia–water mix remains slushy over a much wider range of temperatures, planetary scientists think that it may have played a similar role to volcanic magma on Earth, erupting from fissures and resurfacing areas of many moons. On worlds such as Titan, Pluto and Neptune's satellite Triton, this cold 'cryovolcanism' may still be happening today.

The condensed idea
Several outer solar system moons hide deep oceans

10 Dwarf planets

Only recently recognized as a distinct class of objects, the dwarf planets of our solar system are proving to be one of the most exciting and surprising new territories for planetary exploration. Two in particular – Ceres and Pluto – have been visited by space probes.

At the time when the International Astronomical Union made its historic decision to reclassify the planets in 2006 (see box, opposite), just five objects were granted the newly minted title of dwarf planet: Ceres (the largest member of the asteroid belt) and four Kuiper Belt objects – Pluto, Haumea, Makemake and Eris (in order of increasing distance from the Sun). Hopes that another asteroid, the 525-km (326-mile) Vesta, might meet the criteria were dashed when measurements from the Dawn spacecraft suggested that it does not have sufficient gravity to pull itself into a sphere (even discounting the enormous impact crater at its south pole). Haumea, Makemake and Eris orbit in the depths of the Kuiper Belt, so far away that telescopes can reveal only a few basic facts about them. Fortunately, however, the other two dwarf planets have now been visited by space probes.

THE LARGEST ASTEROID

Ceres was the first asteroid to be discovered, as early as 1801, by Italian astronomer Giuseppe Piazzi. Orbiting between 2.6 and 3.0 AU from the Sun, it lies roughly midway between Mars and Jupiter in the midst of the asteroid belt, and spectroscopic observations from Earth-based telescopes in the 20th century suggested that its surface composition is similar to that of

TIMELINE

1801	1930	2005
Giuseppe Piazzi discovers Ceres, the first asteroid and the innermost dwarf planet	Clyde Tombaugh discovers Pluto – the first Kuiper Belt Object. It is initially classed as a planet	Astronomers discover Eris, an object of similar size to Pluto, orbiting in the scattered disc.

smaller C-type asteroids. These rocky objects are rich in carbonate minerals and are thought to represent material that is essentially unaltered from the early days of the solar system.

However, recent observations reveal a more complex side to Ceres. Images from the Hubble Space Telescope and Keck Telescope showed dark spots on the surface thought to correspond to impact craters, and one surprisingly bright region whose nature would become an enduring mystery. In 2014, with the Dawn mission already en route between Vesta and Ceres, astronomers using the infrared Herschel Space Observatory discovered a thin atmosphere of water vapour being replenished by some form of emission from the surface – most likely the sublimation of frozen surface ice directly into gas. As Dawn approached Ceres in early 2015, it revealed the largest asteroid in unprecedented detail. Its surface proved to be relatively smooth, with a number of low-relief craters. This suggests that Ceres has a soft crust, rich in water ice, which 'relaxes' over time to flatten out elevated or depressed surface features.

Defining dwarf planets

When Pluto was discovered in 1930, it was naturally designated as the solar system's ninth planet. But doubts soon arose about its status, and astronomers began to suspect it was just the first in a hypothetical belt of objects beyond Neptune. Even after further Kuiper Belt Objects were discovered in the 1990s, Pluto clung onto its planetary status – until the discovery of an object designated 2003 UB313, in January 2005. This object, nicknamed Xena and later officially named Eris, had an estimated diameter about 200 kilometres (125 miles) bigger than Pluto, and was promoted by its discoverers as a tenth planet for the solar system.

However, the International Astronomical Union (IAU), responsible for official astronomical nomenclature, had other ideas. Faced with the possibility of many similar worlds lurking in the outer solar system, they appointed a panel of astronomers to come up with an official definition of a planet. Since August 2006, therefore, a planet has been defined as a world in an independent orbit around the Sun, with enough gravity to pull itself into a spherical shape and also to substantially clear its orbit of smaller bodies. The new category of dwarf planet applies to objects that fulfil the first two criteria, but not the last.

2006
The International Astronomical Union introduces a definition of dwarf planets that encompasses Ceres, Pluto and Eris

2015
The Dawn spacecraft enters orbit around Ceres, sending back close-up images for the first time

2015
New Horizons flies past Pluto at high speed, returning a wealth of data

Pluto's moons

Considering its small size, Pluto has a remarkably complex system of moons. The largest, Charon, has just over half the diameter of Pluto itself, and orbits its parent in just 6.4 days. Tidal forces ensure each world keeps the same face permanently turned towards the other. Four smaller bodies, named Styx, Nix, Kerberos and Hydra, orbit slightly beyond Charon.

Left to right: Pluto with its moons Charon, Nix and Hydra, as viewed by the Hubble Space Telescope.

Dawn also discovered numerous bright spots within some craters, one of which seems to be associated with an intermittent haze that appears above it. Chemical analysis of the spots in the months after Dawn's arrival suggested that they may be some kind of salt deposit, but how they accumulate is still uncertain – one theory is that they could be deposited by brine seeping to the surface from a layer of subterranean liquid water.

DEMOTED PLANET

While Dawn was able to enter orbit around Ceres and study it for many months, the New Horizons mission to the outer dwarf planet Pluto was limited to a spectacular high-speed flyby in July 2015. Given Pluto's great distance, the only feasible way of reaching Pluto in a reasonable time frame (a little less than a decade) was to launch a lightweight, high-speed mission on a one-way trip. There was particular pressure to reach Pluto in a hurry while it remained close to the inner edge of its 248-year elliptical orbit – experts suspected that it might develop a tenuous atmosphere while close to the Sun, but this would rapidly freeze to the surface as it retreated from the vicinity of Neptune's orbit into the depths of the Kuiper Belt (see page 44).

Spectroscopic studies in the 1990s had already shown that Pluto's surface is dominated by frozen nitrogen at temperatures of around –229°C (–380°F), with traces of methane and carbon monoxide. The presence of an atmosphere was proved as early as 1985 (detected through tiny changes in the light of distant stars before they pass behind Pluto itself), but atmospheric pressure is little more than a millionth of that on Earth. Unsurprisingly, given that the atmosphere is formed by sublimated surface ice, it too is dominated by nitrogen.

Early attempts to map Pluto in the 1990s and 2000s used the Hubble Space Telescope to monitor a series of mutual eclipses between Pluto and its giant moon Charon. Surface features were impossible to resolve directly, but the variations in brightness and colour caused when each world blocked out part of the other's light revealed strong contrasts in surface brightness and hue. In particular, large dark red patches that are thought to be caused by tholins – complex hydrocarbon molecules formed by methane in the thin atmosphere settling back to the surface.

> **THIS WORLD IS ALIVE. IT HAS WEATHER, IT HAS HAZES IN THE ATMOSPHERE, ACTIVE GEOLOGY.**
>
> Alan Stern, New Horizons Principal Investigator

The biggest surprise from New Horizons' encounter was Pluto's variety of terrain – not only in colour, but also in overall geology. While Ceres is fairly uniform in appearance, Pluto has striking differences that indicate a turbulent geological past and perhaps an active present, also. A bright, heart-shaped area named Tombaugh Regio has a smooth surface with very few craters, and is therefore thought to be relatively young (perhaps a hundred million years old). It seems to be covered by several kilometres of nitrogen ice, and shows features that are unmistakably the work of glaciers. The darker Cthulhu Regio, in stark contrast, is rugged and heavily cratered, and marks one of the tholin patches identified in the Hubble images.

Elsewhere, traces of possible geyser-like gas eruptions have been found, along with a pair of particularly high mountains (c.5 km or 3 miles high) that must be built largely out of water ice. Deep central pits or calderas suggest that these peaks, Wright Mons and Piccard Mons, are cryovolcanoes (see page 39). If confirmed, they will be by far the largest examples so far discovered in the outer solar system.

The condensed idea
The solar system's small worlds can be surprisingly complex

11 Asteroids and comets

The smaller bodies that orbit between the planets can be divided broadly by composition into rocky asteroids and icy comets, although the distinction is somewhat blurred. Alternatively, they can be classified by their orbital regions. This defines asteroid groups, icy centaurs, long and short-period comets, and Kuiper Belt and scattered disc objects.

Following the solar system's formation some 4.6 billion years ago, substantial amounts of material were left in orbit between and beyond the major planets. Jupiter's gravitational influence put an abrupt stop to the growth of Mars and drained planet-forming material from the region close to its orbit (see page 23). This left just a sparse ring of rocky debris, which formed the present-day asteroid belt.

In contrast, beyond the snow line where ice can persist against solar radiation, huge numbers of small icy comets coalesced in orbits that wove between the giant planets. Close encounters were continual, altering the orbits of the planets a little at a time but proving far more traumatic for the smaller bodies. Comets were frequently sent hurtling towards the Sun, or ejected into long, slow orbits up to a light year out. Trillions of comets still linger in this region to this day, forming the Oort Cloud at the outermost limit of the Sun's gravitational influence.

Finally, as Uranus and Neptune moved into their present configuration around 4 billion years ago, they disrupted many of the moderate-sized ice

TIMELINE

1705	1801	1866	1866
Edmond Halley predicts the 76-year orbit of the comet that bears his name	Piazzi discovers the first and largest asteroid, Ceres	Kirkwood identifies gaps in the asteroid belt, confirming that asteroid orbits can evolve over time	Schiaparelli links meteor showers to the orbits of comets

dwarf worlds that had formed around the edge of the solar system. The outer members of this proto-Kuiper Belt remained largely undisturbed and form what is today called the 'classical' Kuiper Belt, but their Sunward cousins were mostly ejected into sharply tilted and highly elliptical orbits, forming the scattered disc.

EVOLVING ASTEROIDS

As the closest of these regions to Earth, the asteroid belt is also the only major reservoir of small bodies to have been found by chance. Following the discovery of Uranus in 1781, many astronomers came to believe in a numerical pattern called Bode's law, which seemed to predict the orbits of the planets, but also a 'missing' world between Mars and Jupiter. In 1801, Italian astronomer Giuseppe Piazzi discovered the largest and brightest asteroid, Ceres, orbiting in this region and many more soon followed.

> SINCE ITS MOVEMENT IS SO SLOW AND RATHER UNIFORM, IT HAS OCCURRED TO ME SEVERAL TIMES THAT [CERES] MIGHT BE SOMETHING BETTER THAN A COMET.
>
> Giuseppe Piazzi

By 1866, enough asteroids were known for US astronomer Daniel Kirkwood to identify a number of gaps in the asteroid belt. These empty regions occur because the orbits of any asteroids within them would bring the space rocks into repeated close encounters with Jupiter. Asteroids falling by chance into such resonant orbits are soon kicked out of it into more elliptical paths. In 1898, the first refugee from these regions, a so-called Near Earth Asteroid catalogued as 433 Eros was discovered by German astronomer Gustav Witt. Several classes of these objects are now known, and their relationship to Earth's orbit is closely monitored as a potential threat.

LONG- AND SHORT-PERIOD COMETS

Similar gravitational interactions with the giant planets also help shepherd the icy bodies of the outer solar system. Comets falling towards the Sun on the relatively brief inner part of their long orbits may find their elliptical

1898	1930	1932	1992
Gustav Witt discovers Eros, the first Near Earth Asteroid	Kenneth Edgeworth and others suggest there is a ring of small bodies orbiting just beyond Neptune	Öpik suggests the existence of a comet cloud surrounding the solar system at a great distance	The Hubble Space Telescope discovers the first Kuiper Belt Object other than Pluto

Sampling the primordial solar system

Asteroids are important to our understanding of the solar system because they retain fragments of material left over from its birth. Based on spectral studies of their light, close-up views from space probe encounters and studies of meteorites (asteroid fragments that fall to Earth), they are divided into several broad groups:

• The C-group of carbonaceous asteroids have dark surfaces and are thought to be rich in unaltered raw materials.
• The S-group of silicaceous or stony bodies show surfaces that have been chemically altered by higher temperatures and geological processes.
• The X-group consist of metallic (mostly iron and nickel) objects.

The S- and X-groups probably originated in relatively large bodies that heated up during formation and whose interiors therefore separated according to density. These objects subsequently broke up in collisions that scattered their fragments throughout the asteroid belt (the belt occupies a vast volume of space, but contains tens of millions of objects, so collisions are frequent on an astronomical timescale). Many asteroid families, united by similarities of composition or orbit, are thought to trace their origins back to such events.

paths radically shortened by an encounter with a giant planet (particularly Jupiter), leaving them with an orbit measured in decades or centuries rather than thousands of years, and an aphelion (furthest point from the Sun) somewhere in the Kuiper Belt. Such short-period comets become frequent and predictable visitors to the inner solar system.

Life close to the Sun dramatically shortens the life expectancy of a comet – each passage around the Sun burns off more of its limited surface ice, and risks a close encounter with Jupiter that could shorten its orbit even more. Some comets end up in rather asteroid-like orbits, taking just a few years to orbit the Sun and rapidly burning through their remaining ice until they deteriorate into dark, dried-out husks that are generally indistinguishable from asteroids.

Comets have been seen by stargazers since prehistoric times, and are easily identified. Their characteristic appearance when close to the Sun is an extended atmosphere, or coma, around a relatively small, solid nucleus, and a tail that always points away from the Sun. English scientist Edmond Halley was famously the first person to calculate the orbital period of a comet in 1705, realizing that the objects seen in 1531, 1607 and 1682 were, in fact, the same body in a 76-year orbit around the Sun. The object in question is now known as Halley's Comet.

DISTANT ORIGINS

The ultimate origins of comets were not clarified until the mid-20th century. The distant Oort Cloud was, in fact, first hypothesized by Estonian astronomer Ernst Öpik in 1932 to explain the fact that long-period comets approach the inner solar system from all directions, and independently put forward by Dutchman Jan Oort in 1950 as a means of explaining how comets could have persisted for the life of the solar system without burning away all of their ices and becoming exhausted.

The Kuiper Belt, by contrast, was proposed by several astronomers in the aftermath of Pluto's discovery in 1930. Dutch-American astronomer Gerard Kuiper's name became attached to it by historical accident, after he wrote a 1951 paper proposing that such a belt might have existed in the *early* days of the solar system. Unlike the Oort Cloud, which can be inferred from various lines of evidence, the existence of the Kuiper Belt was not confirmed until the Hubble Space Telescope discovered 1992 QB1, the first of many new objects that have since been found in the region beyond Neptune.

Comet composition

Germany philosopher Immanuel Kant was the first to suggest that comets were largely made of volatile ice as early as 1755. In 1866, however, Giovanni Schiaparelli linked the annual appearance of showers of meteors (shooting stars) with Earth's passage across the orbits of comets. The idea that comets leave a trail of dusty debris behind them led to a popular model of comet nuclei as floating gravel banks held together by ice. In the early 1950s, however, US astronomer Fred Whipple put forward a 'dirty snowball' theory in which ice is the dominant component. Space probe investigations have subsequently backed up the essentials of Whipple's model, although there are significant variations from one comet to another. In general, they seem to be a mix of carbonaceous dust (including relatively complex organic chemicals) and volatile ices – not only water ice, but also frozen carbon monoxide and dioxide, methane and ammonia.

The condensed idea
Comets and asteroids are the debris of our solar system

12 Life in the solar system?

Could primitive or even relatively advanced forms of life be awaiting discovery amid the myriad worlds of our own solar system? Recent discoveries have revealed not only a surprising variety of potentially viable habitats, but also that life itself can be far more robust than previously thought.

People have speculated on the prospects for life on the other worlds of our solar system since ancient times, but until the late 19th century, when Giovanni Schiaparelli's report of Martian *canali* (see page 28) inspired the first scientific investigation of the subject, alien life remained largely the province of satirists and storytellers. Drawing parallels with Earth, many astronomers were happy to accept that Venus might be a humid, tropical world beneath its clouds, and that the colder and more arid Mars was still capable of supporting primitive plant life, if not the intelligent aliens supposed by Percival Lowell.

From the early 20th century, however, prospects for life suffered a series of dramatic setbacks. In 1926, US astronomer Walter Sydney Adams showed that oxygen and water vapour were almost entirely absent from the Martian atmosphere, and in 1929 Bernard Lyot showed that the atmosphere was dramatically thinner than that on Earth's. Together these discoveries indicated an extremely dry world on which temperatures rarely rose above

TIMELINE

1977	1977	1979
Oceanographers discover flourishing ecosystems around deep-sea vents on Earth	Carl Woese identifies a third kingdom of life, the archaea, that include many extremophile organisms	The discovery of tidal heating raises the chances of liquid water on moons of the outer solar system

freezing, and space probe flybys in the 1960s dealt a death blow to hopes of life on Mars. Early probes to Venus sent back equally grim results – the surface was a toxic furnace that destroyed even heavily shielded landers within minutes.

The subsequent renaissance in the prospects for life in the solar system (and beyond) has emerged from two separate streams of discovery, both of which have gained pace rapidly since the 1970s. In space, probes to distant planets have confirmed that several unexpected worlds harbour large bodies of liquid water that might be suitable for life (most importantly the moons Europa and Enceladus – see Chapter 9), while closer studies of Mars have shown that it may not be quite so arid as previously thought (see Chapter 7).

LIFE AT THE EXTREMES

Just as important, however, are discoveries made on Earth, where a series of breakthroughs have overturned traditional ideas about the conditions within which life can survive and flourish. These began in 1977, when oceanographers using the submersible *Alvin* discovered abundant life around deep-sea volcanic vents on the Pacific Ocean floor. With no sunlight to drive photosynthesis (usually the base of the food pyramid on land and in the oceans), these organisms had instead developed an ecosystem founded on microorganisms that thrive in near-boiling temperatures and digest volcanic sulphur compounds. These bacteria exist in the guts of long tubeworms, and ultimately support other creatures including fish and crustaceans that long ago became isolated on these warm oases in the cold ocean depths.

> **I THINK WE'RE GOING TO HAVE STRONG INDICATIONS OF LIFE BEYOND EARTH WITHIN A DECADE.**
>
> Ellen Stofan, NASA Chief Scientist, 2015

In the late 1970s, US microbiologist Carl Woese investigated the DNA of the deep-sea vent microbes and made the remarkable discovery that they are not simply adapted bacteria, but are instead members of an entirely different

1996	2003	2014
NASA scientists announce possible biogenic molecules and microfossils in a meteorite from Mars	Earth-based astronomers discover signatures of methane in the Martian atmosphere, but follow-up studies are contradictory	NASA's Curiosity Rover detects an abrupt spike in Martian atmospheric methane, probably either volcanic or biogenic in origin

kingdom of life, now known as the Archaea. Distinguished by unique processes in their cell metabolism, archaeans have turned out to be surprisingly widespread in environments ranging from oceans and soils to the human colon. More importantly in the search for alien life, specialized extremophile archaeans also thrive in a range of harsh environments – not only in high and low temperatures, but also extremely arid, salty, acid, alkaline or otherwise toxic conditions.

The evolutionary affinities of the Archaea are still uncertain – they have features in common with both the other major kingdoms of life: the bacteria and multicellular eukaryotes. Some experts think they might be the oldest form of life on Earth, which increases the likelihood that they developed in what we might today regard as an extreme environment. Earth's atmosphere has certainly undergone major changes before reaching its current composition, some of which were driven by the appearance and evolution of life itself. It's certain that the conditions in which the earliest organisms evolved would be inimical to the vast majority of today's life. From the archaean point of view, it is *us* who are the extremophiles.

THE SEARCH FOR LIFE

While the chances that life might have evolved on other worlds have received a significant boost, proving it is a different matter. Currently, our exploration of other planets is limited to robotic probes, and identifying the signatures of past or present life is such a specialized task that few missions have yet been designed with it in mind, and the only one launched so far, the Beagle 2 lander, sadly failed during its landing on Mars in 2003. Mars is the most accessible place to look for life, and the European Space Agency will shortly return to the fray with a two-pronged orbiter/lander mission called ExoMars, specifically designed to look

Panspermia

The idea that life might be seeded from space is an ancient one, but became popular in the 19th century once scientists recognized that material regularly fell to Earth in the form of meteorites. In 1834, Swedish chemist Jöns Jakob Berzelius identified carbon in a meteorite for the first time, and later scientists observed what they thought might be traces of fossilized bacteria inside carbonaceous meteorites. In 1903 another Swede, Svante Arrhenius, suggested that microbes might float through space, driven by the pressure of light from the stars.

More recently, studies of extremophile bacteria and archaeans have shown that microbes can survive in space (especially if sealed inside meteorites) for relatively long times. The discovery of meteorites from both the Moon and Mars has rekindled interest in the idea that life could be transferred between planets in the solar system.

for so-called biosignatures. NASA, meanwhile, is actively developing plans for a future Europa mission, with various concepts also being studied for a targeted investigation of Enceladus. Both spacecraft would be orbiters, but certainly in the case of Enceladus it might be possible to directly detect the signatures of life in material ejected by the moon's famous ice plumes.

Any robotic search for life is inevitably limited in scope compared to what might be achieved by human geologists or biologists, so the final verdict on life in the solar system may

Martian microfossils?

In 1996, a team of NASA researchers made headlines with claims that a meteorite from Mars, catalogued as ALH 84001, contained traces of ancient Martian life. Alongside biogenic molecules, which on Earth would be seen as the work of living organisms, the team found tiny wormlike structures resembling fossils (pictured above). Despite the excitement at the time, other scientists soon raised concerns. Not only did some question whether the biogenic molecules could have entered the meteorite after its arrival on Earth, but one team demonstrated how they could form without the need for life. The alleged microfossils, meanwhile, are smaller than any accepted Earth microorganisms. With so many questions raised, it seems that definitive proof of Martian life will have to await further discoveries.

ultimately have to wait for manned exploration. The identification of meteorites known to have come from Mars (and potentially other worlds) opens up the possibility of a quicker answer, but as the controversy over 'Martian microfossils' shows, such evidence brings complications of its own (see box, above). Indeed, the fact that material can be transferred between worlds in this way raises intriguing questions about the origins of life on our own planet.

The condensed idea
There are viable habitats for life on our cosmic doorstep

13 Our Sun – a star in close-up

The nearest star lies just 150 million kilometres (93 million miles) from Earth, and dominates the solar system. The Sun's proximity means that we can study it in detail and track processes that also take place on most other stars, but which are otherwise impossible to see.

The sharp-edged, incandescent disc that dominates our daytime skies appears at first glance to be the whole of the Sun. But even the earliest astronomers would have seen signs that this was not the case. Most notably, pale but extensive glowing streamers are revealed when the Moon blocks out that blazing disc during a total solar eclipse.

This outer layer of the Sun is called the corona, while the red and pink flame-like loops that arc just above the Moon's dark disc in eclipses are prominences. Around 1605, Johannes Kepler suggested that the corona was produced by tenuous matter around the Sun faintly reflecting its light, but it was not until 1715 that Edmond Halley argued that the Sun had its own atmosphere.

SUNSPOTS

However, it was Galileo's 1612 discovery of dark spots on the solar disc that forever changed our understanding of the Sun's true nature. Sunspots showed that the Sun was not an unchanging sphere, but a mutable,

TIMELINE

1612	1843	1863
Galileo views sunspots for the first time and uses them to measure the Sun's rotation	Samuel Schwabe discovers the periodic variation of sunspot numbers	Carrington measures differential rotation of the Sun, proving it is not a solid body

imperfect, physical object. The motion of the spots allowed Galileo to show that the Sun spun on its axis about once every 25 days.

In the 1760s, Scottish astronomer Alexander Wilson had made a discovery that sent astronomers down a blind alley for the better part of a century. His careful studies of sunspots as they approached the limb (visible edge) of the Sun, showed that they were depressed in comparison to most of the visible surface. This later led William Herschel, who was hugely influential thanks to his discovery of Uranus and numerous other breakthroughs, to conclude that the Sun's bright surface was actually a layer of clouds. These dense clouds shrouded a much cooler solid surface from view, and Herschel speculated that it might even be inhabited. Another German astronomer, Johann Schröter, coined the term 'photosphere' to describe this incandescent apparent surface, and the name stuck.

In the 1870s, however, the theory of a solid Sun was finally discredited by English amateur astronomer Richard Carrington. Through careful measurements, he confirmed that sunspots at different latitudes rotate at different rates. This differential rotation, faster at the equator than at the poles, demonstrated that the Sun was, in fact, a fluid body.

> ONE OF THE GREAT CHALLENGES IN SOLAR PHYSICS IS TO UNDERSTAND, AND ULTIMATELY PREDICT, SOLAR MAGNETIC ACTIVITY.
>
> Dr Giuliana de Toma

THE SOLAR CYCLE
The key discovery that sunspots change in a regular cycle was made in 1843 by Swiss astronomer Heinrich Schwabe, using 17 years of meticulously kept records. Individual spots appeared and died in a matter of days or weeks, but Schwabe identified a cycle in the total numbers that waxed and waned over about ten years. Today this solar cycle is generally agreed to average about 11 years. In 1858, Carrington also showed that sunspots appeared closer to the equator as the cycle progressed.

1908–19	1946	1976
Hale discovers the magnetic nature of sunspots, subsequently using it to explaining the origin of the sunspot cycle	Astronomers observe the solar atmosphere at X-ray and ultraviolet wavelengths for the first time using rocket-borne instruments	John A. Eddy discovers a sustained drop in sunspot numbers around the late 17th century, known as the Maunder Minimum

Cycles on other stars

In general, the magnetic cycles of distant stars make too little difference to their overall light output to be detectable from Earth, but there are exceptions. Flare stars are small and faint red dwarfs that can nevertheless unleash flares far more powerful than any yet seen from the Sun (see page 90). Various techniques can also be used to measure the size and intensity of large sunspots (hundreds of times bigger than those on the surface of the Sun). The simplest technique is called Doppler imaging, and involves measuring slight variations in the star's light output and colour as it rotates. Similar approaches in eclipsing binary stars or stars with transiting exoplanets (see pages 94 and 98) reveal variations in a star's surface as different parts are blocked from view.

More complex methods involve either the Zeeman effect, a modification of absorption lines in a star's spectrum (see page 60) created by intense magnetic fields, or 'line depth ratio', a variation in the intensity of lines that reveals temperature differences on the star's surface. Close monitoring of stars with major spots has revealed stellar cycles similar to our Sun's, but also some that are completely different. For instance, the RS Canum Venaticorum class of variable stars has a cycle in which activity flips from one hemisphere to the other and back again.

The following year brought the first hint that events on the Sun could have a dramatic effect on Earth, when Carrington and others monitored the development of a brilliant spot in the photosphere. Within days, Earth's magnetic field was disrupted by a vast geomagnetic storm affecting everything from the northern lights to the telegraph system. This was the earliest recorded solar flare – a violent eruption of superhot material just above the photosphere – and subsequent studies showed that such events are linked to the same sunspot cycle. They even emanate from the same regions of the Sun (see box).

A MAGNETIC EXPLANATION

In 1908, US astronomer George Ellery Hale discovered that sunspots are regions of intense magnetic fields, and this, along with Carrington's discovery of differential rotation, proved to be key to explaining the solar cycle. The Sun's fluid interior is incapable of generating a permanent magnetic field, but instead an internal layer of swirling, electrically charged hydrogen ions produces a temporary field. At the beginning of a cycle, the field runs smoothly between north and south poles beneath the Sun's surface, but each rotation of the Sun causes it to wind up around the Sun's equator. As field lines become tangled, magnetic loops begin to push out of the photosphere, creating regions of lower density where the heat-transporting mechanism of convection is suppressed (see page 71). As a result, the temperature at each end of these coronal loops is lower and the visible gas appears darker than its surroundings, forming a sunspot. Initially, the magnetic loops push out of the Sun at relatively high latitudes, but as the cycle continues and the field becomes increasingly tangled, the number of loops increases, and they are gradually dragged towards the equator, coinciding with a period of maximum solar activity. The number of solar flares peaks around this time, triggered when field loops short-circuit closer to the Sun's surface, releasing a huge amount of magnetic energy, which heats the surrounding gas to tremendous temperatures and blasts it out across the solar system, still carrying a tangle of magnetic field within it.

Eventually, however, as the sunspots draw closer to the equator, the opposing polarities of the tangled fields begin to cancel out. Loops dwindle in number until eventually the Sun's entire magnetic field has effectively disappeared. After roughly 11 years, this marks the end of the visible sunspot cycle, but it is only the halfway point of the Sun's overall magnetic cycle. A new smooth field is soon recreated beneath the surface, only this time its north–south polarity is reversed, and the entire story repeats again. Only after 22 years does the Sun return to its original magnetic state.

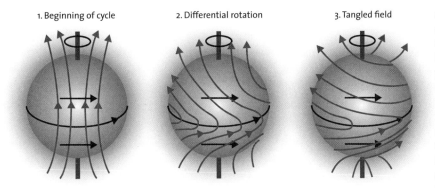

1. Beginning of cycle   2. Differential rotation   3. Tangled field

At the beginning of a solar cycle [1] a weak magnetic field runs beneath the Sun's surface from one pole to the other. As the cycle progresses, the Sun's differential rotation begins to drag the magnetic field around the equator [2]. As the cycle advances, the field becomes more tangled [3] and magnetic loops trigger sunspots and solar flares.

The condensed idea
The Sun's changing magnetism can produce spectacular effects

14 Measuring the stars

Even a casual glance at the night sky reveals variation among the stars – most notably in brightness and colour. Understanding how these differences in appearance reflect physical characteristics reveals that stars are even more varied than their appearance would suggest.

Differences in brightness between stars are an obvious variation that astronomers attempted to catalogue and measure from the earliest times. Around 129 BCE, the Greek astronomer Hipparchus separated stars into six different magnitudes, with the brightest, such as Sirius, in the first magnitude, and the very faintest naked-eye stars in the sixth magnitude. The arrival of the telescope immediately led to the discovery of countless fainter stars, and so the system was extended to higher, fainter magnitudes, but it was not until 1856 that Norman R. Pogson of the Madras Observatory, in what is now Chennai, India, applied a scientific rigour (see box, opposite).

DISTANCE AND LUMINOSITY

Do differences in the 'apparent' magnitude of stars reflect variations in their inherent luminosity, their distance from Earth, or a mixture of the two? William Herschel, in his first attempt to map the Milky Way (see page 136), mistakenly assumed that all stars were more or less as bright as each other, and so magnitude was an indication of proximity to the Earth. The question remained unresolved until 1838, when German astronomer Friedrich Bessel successfully measured the distance to one of the closest stars, a binary pair called 61 Cygni, using its shift in parallax (see box on page 58): it turned out

TIMELINE

1827	1838	1856
Félix Savary calculates the orbit of the binary star Xi Ursae Majoris, key to determining its mass	Bessel successfully measures the distance to 61 Cygni using stellar parallax	Pogson formalizes the system of apparent magnitude used for measuring the brightness of stars

to lie roughly 100 million *million* kilometres (60 trillion miles) from Earth – a distance so vast that it is easier to refer to it in terms of the amount of time its light takes to reach us: hence, 10.3 'light years'. At this distance, the magnitudes of its stars – 5.2 and 6.05 – implied that they were 1/6 and 1/11 as luminous as the Sun, respectively.

As technological improvements allowed the direct measurement of more stellar distances later in the 19th century, it soon became clear that the stars were extremely varied in terms of brightness. Sirius, for instance, lies on our cosmic doorstep at just 8.6 light years away, and is about 25 times more luminous than the Sun. Canopus, the second brightest star in the sky, is an estimated 310 light years away (too far to measure using Bessel's method until recently), and must therefore be more than 15,000 times as luminous as our own star.

The modern magnitude system

Based on careful comparisons between stars, 19th-century astronomer Norman R. Pogson estimated that a first-magnitude star was about 100 times brighter than a fifth-magnitude one, and suggested standardizing this relationship so that any difference of exactly 5 magnitudes corresponds to a factor of 100 difference in brightness (a magnitude difference of 1.0 therefore corresponds to a brightness difference of 2.512 times, known as Pogson's ratio). Pogson set the magnitude of the north pole star Polaris as exactly 2.0, and as a result found that the magnitude of the very brightest stars was pushed into negative values (so that Sirius's magnitude is officially –1.46). After Pogson's time, astronomers discovered that Polaris was very slightly variable, so the modern reference point is the bright star Vega (Alpha Lyrae), defined as magnitude 0.0.

Even without direct distance measurements, however, astronomers can sometimes take a shortcut to determine relative stellar luminosities. This relies on the assumption that stars in tight clusters such as the Pleiades of Taurus (which are too closely grouped to be a chance statistical alignment) are all effectively at the same distance from our solar system. Differences in apparent magnitude therefore reflect differences in 'absolute' magnitude or luminosity.

1869

Gustav Kirchoff quantifies the relationship between the colour and surface temperature of stars

1989–93

The European Space Agency's Hipparcos mission carries out the first large-scale parallax survey from space

The parallax method

The only direct way of measuring the distance to a star uses parallax – the shift in a nearby object's position against a more distant background caused when the observer's point of view changes. Once it was understood that Earth orbited the Sun, and the true scale of the solar system was appreciated, the shift in Earth's position from one side of its orbit to the other (approximately 300 million kilometres or 186,000 miles) provided an ideal baseline for such measurements, though only the nearest stars would show sufficient parallax shift to be measured with 19th-century technology. Potential targets were chosen on the basis of their high proper motions, or movements across the sky (see page 65). However, it still took many years of effort for Friedrich Bessel to achieve his measurement of 61 Cygni's parallax – a mere 0.313 second of arc or 1/11,500th of a degree. Today, satellites such as the European Space Agency's Gaia can measure angles 50,000 times smaller than this.

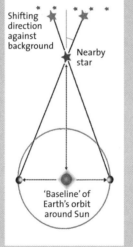

Shifting direction against background

Nearby star

'Baseline' of Earth's orbit around Sun

COLOUR, TEMPERATURE AND SIZE

But luminosity is not the whole story. Another important stellar property is colour. Stars turn out to have a whole range of hues from red and orange through yellow to white and blue (though, curiously, only one star in the sky is generally agreed to be green). There's an intuitive link between such colours and the ones emitted by, for example, an iron bar heated in a furnace, but the relationship between temperature and overall colour was only formalized by Gustav Kirchhoff in 1869. Kirchhoff identified a characteristic radiation curve that describes the amount of radiation of different wavelengths and colours emitted by a 'black body' of a given temperature (a black body is a hypothetical object that absorbs light perfectly, but stars turn out to behave very similarly). He found the hotter the object, the shorter-wavelength and bluer its overall radiation. This meant that, with the development of stellar spectroscopy around the turn of the 20th century (see page 60), it became possible to estimate a star's actual surface temperature from its colour. This new tool was surprisingly powerful, since it allowed astronomers to estimate the *size* of stars for the first time.

The principle behind such an estimate is fairly straightforward. First, calculate the power output needed to heat a square metre of the star's surface to the measured temperature (using a simple equation called the Stefan–Boltzmann law). Then work out the star's *overall* power output or luminosity (by comparing its distance and apparent magnitude). The star's surface area is then easily calculated, and this is, in turn, dependent on its diameter.

To give a concrete example, a relatively small yellow star like the Sun has an average temperature of 5,800°C (10,472 °F) as a result of its internal power output (of 1 solar luminosity) heating its surface. In contrast, the unstable yellow star Rho Cassiopeiae goes through phases when it has a similar surface temperature to the Sun, despite being an incredible half a million times more luminous (implied by a distance of about 8,200 light years and an apparent magnitude of 6.2). This means that its diameter must be about 500 times that of the Sun. It is in fact a yellow hypergiant (see Chapter 29), a star so large that in our solar system it would extend past the orbit of Mars.

TO ACCOMPLISH THIS HAS BEEN THE OBJECT OF EVERY ASTRONOMER'S HIGHEST ASPIRATIONS...
John Herschel, on Bessel's measurement of stellar parallax

WEIGHING STARS

One final key stellar property is mass, but how can we weigh a star? Until recently the only way to directly measure stellar masses has been by calculating the orbits of binary systems (see page 95). The stars in such systems orbit around a shared centre of mass called the barycentre, at average distances determined by their relative masses (so the more massive one sits closer to the barycentre). French mathematician and astronomer Félix Savary worked out the first binary orbit in this way as early as 1827. When combined with information from spectroscopic binaries or parallax measurements, it's possible to find more detailed parameters for the orbits of certain binaries. Then one can calculate either the exact masses or range of masses involved, but even a knowledge of the *relative* masses has proved invaluable to understanding the evolution of stars (see page 76).

The condensed idea
The colour and brightness of a star reveal its distance and size

15 Stellar chemistry

Spectroscopy is a technique for discovering the chemical constituency of materials from the light they emit. It has a huge variety of applications in chemistry and physics, but is particularly important to astronomy, where the light from distant objects is usually our only means of studying them.

In 1835, French philosopher Auguste Comte declared that, when it came to the stars, 'we would never know how to study by any means their chemical composition'. The next few decades would prove him spectacularly wrong, but it seems unfair to criticize him for lack of foresight – plenty of others also overlooked evidence that had been discovered more than 20 years previously.

From 1814 onwards, the German optician Joseph von Fraunhofer had published details of discoveries with his new optical inventions: the spectroscope and diffraction grating. Both instruments could study the spectrum of sunlight far more precisely than by simply splitting it with a glass prism. Fraunhofer found that the solar spectrum, far from being the continuous rainbow that Isaac Newton had identified more than a century before, was actually riddled with narrow, dark lines, indicating that specific colours of light were being blocked by unknown substances. Fraunhofer mapped some 574 lines in the solar spectrum, and even found dark lines in the spectra of bright stars such as Sirius, Betelgeuse and Pollux. He also showed that some stellar lines matched those in the Sun, while others differed.

TIMELINE

1814	1842	1848	1859
Fraunhofer discovers dark lines in the solar spectrum	Doppler describes the shift in the wavelength of light caused by relative motion of source and observer	Hippolyte Fizeau suggests the Doppler effect will show itself most clearly in the shifting of spectral lines	Kirchoff and Bunsen link spectral lines to the presence of particular elements

ELEMENTAL FINGERPRINTS

The origin of the so-called Fraunhofer lines remained unclear until 1859, when German chemists Gustav Kirchhoff and Robert Bunsen linked them to atoms in the solar atmosphere. Kirchhoff and Bunsen had been using the spectroscope to investigate the colours of light produced when various substances burned in a flame. They found that these tended to be a mix of a few very specific colours, and each element produced a unique bright line spectrum. Realizing that the colours of light emitted by burning substances corresponded to some of the dark lines in the solar spectrum, they inferred that they were caused by the same elements absorbing light.

The full explanation for the origin of what are now called absorption and emission spectra had to wait until the early 20th century, when Danish physicist Niels Bohr described how they result from the configuration of electron particles at different energy levels within an atom. When bombarded by a broad range of light (a continuum spectrum), such as the black-body emission from the surface of a star (see page 58), electrons absorb the specific frequencies that allow them to briefly jump to higher energy levels. Since each element has a unique electron configuration, it creates a unique pattern of absorption lines. Emission spectra, meanwhile, are created when energized electrons drop back to a more stable, lower

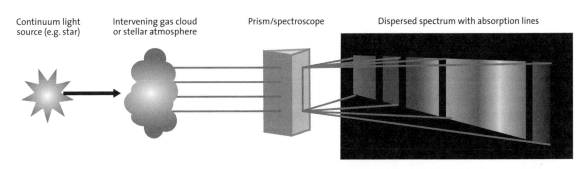

Continuum light source (e.g. star)   Intervening gas cloud or stellar atmosphere   Prism/spectroscope   Dispersed spectrum with absorption lines

1866	1868	1890	1913
Secchi develops the first system of stellar classification based on spectral lines	Huggins uses the Doppler effect on spectral lines to determine the speed of a star's motion away from Earth	First edition of the Henry Draper Catalogue is published	Bohr explains how changes in the energy state of atoms give rise to spectral lines

energy level, getting rid of their excess energy as a small packet of light (a photon) with its own specific wavelength, and hence, colour.

In the aftermath of Kirchhoff and Bunsen's breakthrough, astronomers looked again at the Fraunhofer lines and successfully linked them to elements such as hydrogen, oxygen, sodium and magnesium, present in the Sun's outer layers. In 1868, French astronomer Jules Janssen and Briton Norman Lockyer independently identified lines in the solar spectrum that could not be linked to any known element. Lockyer concluded that the Sun contained a significant element so far undiscovered on Earth, and named it helium after Helios, the Greek sun god.

The Doppler effect

The presence of absorption lines in starlight provides a very convenient set of markers for measuring a star's motion thanks to the Doppler effect. This is a shift in the frequency and wavelength of waves reaching an observer that depends on the relative motion of the wave source. It was first proposed by Austrian physicist Christian Doppler in 1842, who hoped it might explain the different colours of starlight: light from stars moving towards us has higher frequencies and shorter, bluer wavelengths, while light from stars moving away from us has lower frequency, longer wavelengths and appears reddened. Alas for Doppler, the high speed of light makes the effect much weaker than he had anticipated (in all but the most extreme circumstances – see page 160) – but it was confirmed in sound waves in 1845. The Doppler effect is not the explanation for the colours of stars, but 'red shifts' and 'blue shifts' of absorption lines from their expected positions can be used to precisely measure the speed of an object's motion towards or away from Earth. William Huggins was among the first to attempt this for stars, but it was Angelo Secchi and German astronomer Hermann Vogel, in the 1870s, who successfully used Doppler shifts in absorption lines on different parts of the Sun in order to demonstrate its rotation.

SPECTRAL CLASSIFICATION

Other astronomers focused on the spectra of the stars, and two of the most productive were William Huggins in London, and Angelo Secchi in Rome. Secchi developed a basic classification system for spectra, identifying four main classes of stars: those with spectra similar to the Sun; blue-white stars with spectra like Sirius; red stars with broad absorption bands like Betelgeuse; and so-called carbon stars (also typically red, but with strong carbon absorption lines).

Huggins, meanwhile, was the first to realize that the light emitted by the diffuse objects known as nebulae consisted of just a few precise emission lines, and correctly concluded that they are enormous clouds of hot, energized interstellar gas. Huggins was a pioneer of astrophotography and went on to create some of the first extensive photographic catalogues of stellar spectra. His work, however, was eclipsed by the efforts and later legacy of Henry Draper, a US doctor and amateur astronomer who captured the first photographs of both stellar and nebular spectra before succumbing to pleurisy at the age of just 45 in 1882. In 1886, Draper's widow Mary Anna donated money and equipment to Harvard College Observatory to fund the most ambitious astronomical project of the age: a large-scale photographic catalogue of stellar spectra. Known as the Henry Draper Catalogue, it took almost four decades to complete and ultimately described the spectra of more than 225,000 stars.

> **THE PATH IS OPENED FOR THE DETERMINATION OF THE CHEMICAL COMPOSITION OF THE SUN AND THE FIXED STARS.**
> Robert Bunsen

The driving force behind the catalogue was observatory director Edward Pickering, but the bulk of the work was carried out by a team of women known to history as the Harvard Computers. Pickering's motivation for hiring a female team was partly driven by budgetary issues: women would work for lower wages than men and so he could afford a bigger team to analyse the huge amounts of data generated by his photographic survey. Many of his team, however, proved to have impressive scientific talents, and were responsible for several important breakthroughs in the way we understand the properties of stars.

The bulk of the initial cataloguing work fell to Pickering's first recruit and former maid, Scots-born Williamina Fleming. She extended Secchi's classification system, assigning each star a simple letter from A to N depending on the strength of hydrogen lines in its spectrum (with O, P and Q applied to objects with unusual spectra). This system, used in the first Draper Catalogue published in 1890, would undergo several major changes before becoming the classification we use today.

The condensed idea
Starlight bears the fingerprints of chemical composition

16 The Hertzsprung-Russell diagram

Perhaps the most important breakthrough in understanding the life cycles of stars came when early 20th-century astronomers compared the newly catalogued spectral types of stars with their luminosities. The resulting plot of stellar properties, called the Hertzsprung-Russell (H-R) diagram, changed astronomy forever.

The first fruit of William Pickering's research project at Harvard College Observatory (see page 63) was the Draper Catalogue of Stellar Spectra, published in 1890. Mostly compiled by Williamina Fleming, it contained spectra for some 10,351 bright stars. While work continued to add more stars to the main catalogue, Pickering and his all-female team of 'computers' investigated some of the brighter spectra in greater detail.

MAURY'S SYSTEM
Among the most talented of the Harvard Computers was Antonia Maury, a niece of Henry Draper. She began to notice significant features in the spectra of brighter stars that had been overlooked in Fleming's simple alphabetical classification. Not only did the spectral lines vary from star to star (indicating different elements in their atmosphere), but the intensity and width of the lines varied between stars with apparently identical chemistry. Believing that line width represented something fundamental about the nature of the stars, Maury proposed a reordering of the spectral

TIMELINE

1890	1890s	1901
First edition of the Henry Draper Catalogue is published	Maury pioneers classification of stars based on the width of spectral lines	Cannon devises the final version of the Harvard Classification Scheme for spectra

types to reflect their strength. Pickering and Fleming, however, both found the new classification system to be overly complex, and Maury ultimately quit the project. Despite Pickering's requests, though, she refused to give up her work on spectral line widths, and held out for official acknowledgement when her catalogue of some 600 stars was finally published in 1897.

While Pickering continued to downplay the significance of Maury's ideas, they influenced her successor. Annie Jump Cannon had joined the Harvard Group to study the southern-hemisphere stars now being added to the catalogue. She introduced her own classification system, which combined the simplicity of Fleming's letters with Maury's line-width approach. Dropping several letters and reordering them to reflect spectral colours from blue to red resulted in a sequence of spectral types O, B, A, F, G, K and M.

SPECTRA FORM THE KEY

A few years later, the mystery of the line-width variations was taken up by Danish astronomer Ejnar Hertzsprung. He used an ingenious rule of thumb to estimate the distance, and therefore brightness, of stars that could not be measured directly through the parallax method. As a general rule, he argued, more distant stars would display smaller proper motions (year-on-year movements across the sky, caused by the relative motion of the star and our solar system). Proper motion could therefore be used as a crude proxy for distance – if two stars showed the same apparent magnitude, one could guess that the one with the smaller proper motion was more distant and therefore had greater inherent luminosity.

> CLASSIFYING THE STARS HAS HELPED MATERIALLY IN ALL STUDIES OF THE STRUCTURE OF THE UNIVERSE.
> Annie Jump Cannon

Using this method, Hertzsprung identified a broad division among stars of similar colours, separating luminous giants from more numerous but fainter dwarfs, particularly at the cooler end of the spectrum. He now discovered

1908	1911	1913
Hertzsprung links Maury's line width variations to the intrinsic luminosity of stars	Hertzsprung publishes a basic form of the H–R diagram for stars in the Pleiades	Russell produces the first H–R diagram to include the full variety of stars

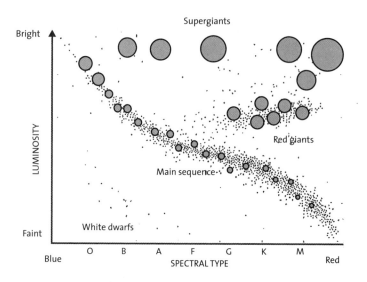

that stars with narrow spectral lines were more luminous than those with broad lines, and backed up this hypothesis by laboriously calculating the distance to several groups of stars. (The reason for the difference in line width eventually became apparent a few years later – see page 119.)

In 1911, Hertzsprung published a chart comparing the spectral features of stars in the Pleiades cluster (a proxy for their surface temperature and Annie Jump Cannon's spectral type) with their apparent magnitudes (a reflection of their absolute magnitudes, since all the cluster stars lie at the same distance). Though necessarily limited because the stars it represents are all broadly similar, the chart showed an unmistakable trend – the more luminous a star is, the hotter its surface.

EXPANDING THE DIAGRAM

Over the next couple of years, Henry Norris Russell, working at Princeton University, developed Hertzsprung's work with a far more ambitious chart based on the same idea. Russell's diagram showed a much broader range of stars, including those from the Hyades cluster (an older and more varied group), and those whose luminosities could be accurately worked out from parallax measurements. It compared spectral type to estimated absolute magnitude, revealing some fundamental patterns for the first time.

The vast majority of stars lay on a diagonal strip ranging from cool and red, to hot and blue. This band, previously identified by Hertzsprung, encompassed all of his dwarfs, and was named the main sequence. Far less common giants and supergiants were scattered across the top of the table in all colours and temperatures, with a strong concentration of luminous red and orange giants emerging in a branch from the main sequence.

The Hertzsprung–Russell diagram proved hugely influential, and over the next two decades astronomers continued to add to it. Certain types of variable star (see page 112) always fell into specific areas of the chart, while new types of star were discovered that filled gaps (see page 124). The fact that overwhelming numbers of stars lay on the main sequence showed that this is where the vast majority of stars spend most of their lives. The entire approach could also be inverted – close analysis of a star's spectrum alone could reveal its place on the diagram, giving not only a measurement of its spectral type and surface temperature, but also a rough idea of its intrinsic luminosity, and therefore its distance from Earth.

Distances and the H–R diagram

The establishment of the H–R diagram made it possible to get a rough idea of a star's intrinsic luminosity (and therefore its distance) simply from its spectral properties. But the main sequence and other regions can be so broad that deriving an individual star's luminosity from its spectral type alone always involves a certain amount of guesswork. Fortunately, the H–R diagram also permits a far more accurate measurement of the distance to star clusters – a technique called main sequence fitting.

Since all stars in a given cluster are effectively at the same distance from Earth, differences in their apparent magnitude are a direct reflection of differences in their absolute magnitudes. This makes it possible to plot an H–R diagram for a specific cluster, which should show the same characteristic main-sequence distribution as the generalized version. Finding the difference between observed and intrinsic brightness is then simply a matter of calculating the offset between the two graphs, and since many stars are involved, it's possible to do this with high precision.

Of course, as with most astronomical techniques there are some complicating factors – for example, the proportion of heavy elements in a cluster's stars affects their distribution somewhat. Also, as a cluster gets older its more massive stars start to leave the main sequence, so it's important to pin down the actual main-sequence stars with precision and exclude stragglers.

The condensed idea
Comparing colour and brightness reveals the secrets of the stars

17 The structure of stars

Understanding the internal structure of stars is key to explaining the differences between them. However, with the development of the Hertzsprung–Russell diagram in the early 20th century, astrophysicists began to properly appreciate just how varied stars can be.

Despite the breakthroughs in stellar spectroscopy around the turn of the 20th century, astronomers knew surprisingly little about the internal composition of stars. Spectral lines were assumed to explain only the constituents of the atmosphere, and the question of what lay beneath the photosphere remained unanswered. However, English astronomer Arthur Eddington showed that it was possible to develop a sophisticated model of stellar interiors without reference to the exact elements present. Eddington had garnered an international reputation in 1919, when he provided experimental proof of Albert Einstein's General Theory of Relativity (see page 191). He had also been investigating the structure of stars, and in 1926 he published his hugely influential book, *The Internal Constitution of Stars*.

BALANCED LAYERS

Eddington's approach, based on the fact that temperatures inside the Sun were clearly hot enough to melt any known element, was to treat the interior of a star as a fluid, caught between the inward pull of gravity

TIMELINE

1906	1925	1926
Schwarzschild investigates the balance between the thermal pressure of stars and the inward pull of gravity	Cecilia Payne argues that the Sun is predominantly made of hydrogen	Eddington's *The Internal Constitution of Stars* introduces the idea of outward radiation pressure from the core

and the external force of its own pressure. Astronomers such as Germany's Karl Schwarzschild had already investigated this idea using models that assumed the outward pressure was entirely due to thermal factors, but had met with mixed results. Eddington, however, thought that, as well as the pressure caused by hot atoms ricocheting around with huge amounts of kinetic energy, an effect called radiation pressure also had a role to play. According to his theory, radiation was being generated in the star's core (rather than entirely at its surface, as most astronomers then believed). This exerted a substantial pressure of its own as its photons pinged off individual particles at different depths within the Sun.

> **AT FIRST SIGHT IT WOULD SEEM THAT THE DEEP INTERIOR OF THE SUN AND STARS IS LESS ACCESSIBLE TO... INVESTIGATION THAN ANY OTHER REGION OF THE UNIVERSE.**
> Arthur Eddington

Eddington was forced to work out much of his theory from first principles, but concluded that stars could only remain stable if energy generation happened entirely in the core, at temperatures of millions of degrees (far greater than the surface of even the hottest star). He showed that thanks to the thinning out of radiation at greater distances from the core, any arbitrary layer in the star would be held in hydrostatic equilibrium. In other words, at every point in the star, the outward radiation and thermal pressure were just enough to counterbalance the inward force of gravity.

The internal structure of a star, Eddington argued, was governed by changes in the opacity of its materials. Another British astronomer, James Jeans, had argued that at such high temperatures, atoms would be entirely ionized (stripped of their electrons and reduced to bare atomic nuclei), and Eddington realized that different degrees of ionization (at different temperature and pressure levels within the star) would affect whether the interior was opaque or transparent. Eddington's theory of stellar interiors proved successful in predicting how stars behave: most notably, it provided an explanation for stars that pulsate in periodic cycles (see page 112).

1930	1938	1975
Unsöld discovers that material in the outer parts of Sunlike stars forms a convection zone	Öpik argues against the general assumption that stars are well mixed	Gough suggests using helioseismology to probe the internal structure of the Sun

Energy in main sequence stars can be transported from the core to the surface of stars by either convection or radiation, but the depth and location of the different transport zones varies depending on the star's mass.

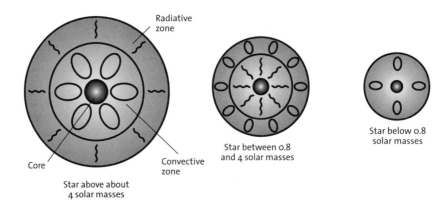

Radiative zone

Core

Convective zone

Star above about 4 solar masses

Star between 0.8 and 4 solar masses

Star below 0.8 solar masses

THE HYDROGEN BREAKTHROUGH

Setting aside the question of chemistry allowed Eddington to think about stellar structure in powerful abstract terms, but such details would be necessary to understand both how stars shine, and how they evolve and change structure over time (see pages 72 and 76). By coincidence, just as Eddington was writing his book in Cambridge, Cecilia Payne was working on her doctoral thesis at Harvard. In it, she made a key breakthrough linking the presence and strength of spectral lines to the temperature of a star's photosphere. This provided the key to identifying the elemental composition of the Sun.

The intensity of spectral lines had previously been explained as a direct indication of the relative abundances of elements in a star's atmosphere, but Payne showed that they were in fact mostly due to temperature differences. Using this new approach, she calculated that the ratios of oxygen, silicon and carbon in the Sun's atmosphere were broadly similar to those on Earth, but also found that our star contained far greater amounts of helium and especially hydrogen than anyone had previously suspected. Payne concluded that these two elements were the dominant constituents of the Sun, and indeed of all stars. It took several years before her idea was widely accepted.

ENERGY TRANSPORT ZONES

One crucial area where Eddington got it wrong was in his assumption that a star's interior would be homogeneous, with the same elemental composition throughout. He thought that energy production in the hottest central region would cause the entire star to churn. Convection currents would carry hotter material upwards and cooler material would sink downwards, ensuring that the interior was thoroughly mixed.

According to theories first put forward by Ernst Öpik in 1938, however, material circulating in a star's core remains there throughout its history. The core is surrounded by a deep radiative zone, where high-energy photons rebound off particles generating enormous pressures. In Sun-like stars this is capped by a further layer of relatively cool convecting material whose existence was confirmed by German astrophysicist Albrecht Unsöld in 1930. This change in the method of energy transport is caused by a transition where the cooler material suddenly becomes opaque. At the top of the zone, the Sun becomes transparent again, but the particles here are far less densely packed, so radiation emitted from the rising gas can simply escape into space. This is what forms the star's incandescent surface or photosphere, which we see from the Earth.

Helioseismology

The most direct way of studying the structure of the Sun or any star is to use the sound waves that constantly ripple through it. These seismic waves are analogous to the ones that cause earthquakes on our own planet. In 1962, solar physicists from the California Institute of Technology using spectroscopy to study the Sun discovered an oscillating pattern of cells, each around 30,000 km (19,000 miles) in diameter shifting up and down in a roughly 5-minute period. The cells were assumed to be a surface effect until 1970, when another American solar physicist, Roger Ulrich, suggested they might be a standing pattern caused by waves oscillating back and forth in the Sun's interior. A few years later in 1975, Douglas Gough demonstrated how Ulrich's oscillating 'p-waves' could be used to probe the Sun's interior. From the way they affected the surface patterns, Gough identified boundaries within the Sun, such as that between the convective and radiative zones.

The condensed idea
Inside stars, gravity and pressure are finely balanced

18 The power source of stars

The question of just how the Sun and other stars generate their light and heat was a longstanding mystery in astronomy, and one that could not be satisfactorily solved by classical physics alone. The stellar energy puzzle could only be resolved with the arrival of nuclear physics in the 20th century.

The earliest theories attempting to explain the Sun as a physical object assumed that our star was nothing more sophisticated than an enormous ball of coal or some other combustible substance, burning merrily in space. The chemistry of combustion was poorly understood, as was the lack of oxygen in space, and so it was not until 1843 that Scottish astronomer John James Waterston produced a proper analysis of the implications. He showed that, if the Sun shone with its current intensity throughout its history, it would only contain sufficient material to burn for about 20,000 years, even if the chemical reaction was extremely efficient.

GRAVITATIONAL POWER

Scientists at the time had little idea of the true age of the Earth and solar system, but geological and fossil discoveries were already beginning to show that an age of many millions of years was more likely than the few thousand years widely inferred from the Bible, so the search was on for a new mechanism to power the Sun. Waterston himself suggested that energy

TIMELINE

1843	1854	1856–1890s
Waterston demonstrates that the Sun cannot be powered by a chemical reaction such as combustion	Helmholtz proposes a mechanism that would allow stars to generate energy from gravitational contraction, later modified by Kelvin	Various estimates put the Sun's lifespan under the Kelvin–Helmholtz mechanism at around 20 million years

might be released by a constant infall of small meteors onto the surface, but a more plausible theory was proposed in 1854 by German physicist Hermann von Helmholtz. He argued that the Sun's energy arose from the effects of its own gravity causing it to shrink and heat up over time. With modifications from British scientist Lord Kelvin, this 'Kelvin–Helmholtz mechanism' offered a way for the Sun to generate energy at its present level for more than 100 million years. This dovetailed fairly neatly with ideas about the age of the Earth, which geologists believed could only be tens of millions of years old, since otherwise its interior would have cooled and solidified.

The gravitational theory began to fall apart around the turn of the 20th century, when the discovery of new radioactive elements revealed a way of keeping the Earth's interior hot for much longer. Darwin's theory of evolution, meanwhile, suggested that the variety of current life would have taken many hundreds of millions, if not billions, of years to arise through natural selection. So by the time Arthur Eddington turned his attention to the problem in his 1926 masterwork on stellar structure (see page 68), the question of the Sun's energy source was open once again.

> **PROBABLY THE SIMPLEST HYPOTHESIS... IS THAT THERE MAY BE A SLOW PROCESS OF ANNIHILATION OF MATTER.**
> Arthur Eddington

ENERGY FROM MASS

Eddington calculated that gravitational contraction would cause some stars to show dramatic changes on the kind of century-long timescales covered by astronomical records. Since there are no such changes, the power source must be much longer-lived and more stable. He also dismissed the meteoric impact theory, since it would be unable to influence processes in the heart of a star. Instead, he argued, the only plausible power source of stars was subatomic in nature: when mass was rendered into energy by Einstein's famous equation $E = mc^2$, a star like the Sun turned out to contain more than enough matter to shine through a multi-billion-year life cycle.

1926	1927	1937	1939
Eddington suggests the Sun is powered by nuclear reaction that converts mass directly into energy	Arthur Holmes publishes evidence that the Earth is several billion years old	Gamow and Weizsäscker outline the proton-proton fusion chain that is the main source of power for Sun-like stars	Bethe discovers the CNO cycle that plays a major role in stars more massive than the Sun

The proton–proton chain

The proton–proton chain involves the fusion of two hydrogen nuclei (protons), one of which spontaneously transforms into a neutron to create a stable deuterium (heavy hydrogen) nucleus. Fusion with another neutron creates another stable isotope, helium-3, and finally two helium-3 nuclei fuse to create normal helium-4, releasing two 'spare' protons in the process. Energy is released in increasing amounts at each stage of the process, and Bethe also recognized other branches the reaction could take, generally in stars with hotter interiors than the Sun (see The CNO cycle, opposite).

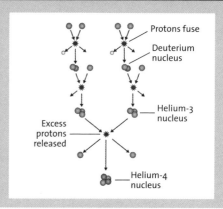

But how was this energy liberated? Eddington considered three main options – the radioactive breakdown of heavy atomic nuclei (fission), the combination of light nuclei to make heavier ones (fusion) and the hypothetical 'cancellation' of matter, when oppositely charged electrons and protons were forced together. He soon concluded that fusion was the most likely mechanism. By way of demonstration, he pointed out that a helium nucleus had 0.8 per cent less mass than the four hydrogen nuclei required to create it (a 'mass defect' that represents the mass released as energy during fusion). As Cecilia Payne's ideas about stellar composition (see page 70) were accepted from the late 1920s, astronomers realized that hydrogen and helium were indeed the dominant elements within stars.

BUILDING THE FUSION MODEL

One major problem for Eddington's fusion theory was that temperatures in the Sun did not seem high enough to support it. Positively charged particles are mutually repelled, so temperatures would need to be exceedingly high for individual protons to collide and merge. In 1928, however, Russian physicist George Gamow applied the strange new science of quantum mechanics to show how protons could overcome this repulsion and fuse together. By 1937, he and his German colleague Carl Friedrich von Weizsäcker were able to propose a proton–proton chain in which collisions between hydrogen nuclei gradually created helium – the very idea that Eddington had toyed with a decade previously.

Gamow and Weizsäcker's chain had problems of its own, namely that it involved the production of highly unstable isotopes (atomic variants) that would disintegrate the moment they formed, rather than sticking around long enough to join with further protons and create stable helium. In 1938, Gamow invited a small group of leading nuclear physicists to a conference in Washington to discuss the problem, including German émigré Hans Bethe. Bethe initially had little interest in the problem, but intuitively saw a possible solution, and rapidly worked out the details with Charles Critchfield. The following year he published two papers outlining not only the process of hydrogen fusion that dominates in Sun-like stars, but also an alternative process called the CNO cycle that takes place mostly in the hotter interiors of more massive stars (see box, right). By rigorously analysing the rate at which the two processes take place in various conditions, Bethe was able to explain not only how stars shine, but also how their various fusion processes give rise to a variety of relatively heavy elements.

The CNO cycle

In conditions hotter than the core of our Sun, carbon can act as a catalyst, speeding up the rate at which hydrogen is fused into helium, while itself remaining unaltered. Hydrogen nuclei (protons) fuse with the carbon nucleus to create nitrogen and then oxygen. Finally, when a further proton attempts to fuse with the oxygen nucleus, it disintegrates, releasing a fully formed helium nucleus and restoring the original carbon. Once again, energy is released at all stages of the process. The CNO cycle becomes dominant in stars with more than 1.3 solar masses, and is so fast and efficient that its presence or absence in stars is a key factor in determining their lifespans (see page 77).

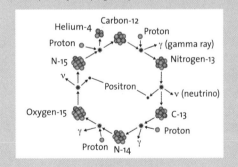

The condensed idea
Stars shine by fusing atomic nuclei to release energy

19 The life cycle of stars

Together, the Hertzsprung–Russell diagram and breakthroughs in the understanding of stellar power sources finally helped astronomers get to grips with one of the biggest scientific mysteries of all – how stars live and die. However, the journey involved abandoning some widely held theories.

Henry Norris Russell's 1912 presentation of the first diagram plotting luminosity against spectral type of stars raised huge questions for astrophysicists. The huge numbers of stars found along the diagonal main sequence between dim red and brilliant blue stars clearly implied that this was where the vast majority of stars spent most of their lives, but how should the patterns be interpreted? The diagram's pioneers had differing viewpoints, but both believed the graph reflected a star's evolution. Russell suspected that stars began life as red giants, contracted to become bright blue stars, and then slowly faded, moving down the main sequence and cooling as they aged. Ejnar Hertzsprung's ideas were less specific, but he thought the main sequence and the horizontal band of multicoloured giants and supergiants along the top of the diagram represented two different evolutionary tracks.

THE MASS-LUMINOSITY RELATIONSHIP

Various theories jostled for position until the mid-1920s, generally rooted in the idea that stars were powered by some form of gravitational contraction (see page 72). Arthur Eddington's 1926 book on the structure of stars, however, brought a fresh approach to proceedings. Working from

TIMELINE

1913	1926	1938
Astronomers initially interpret the main sequence of an H-R diagram as an evolutionary track	Eddington highlights the importance of the mass-luminosity relationship to stellar evolution	Öpik argues that stellar material is not well mixed, limiting the fuel supply and therefore the age of stars

his theoretical model of stellar interiors, he calculated that there was a fundamental relationship between mass and luminosity for nearly all stars: the more massive the star, the brighter it should be.

This idea was not new, and Hertzsprung himself had previously found some evidence for it in binary stars (see Chapter 23). Eddington's approach, though, confirmed theoretically that mass increased with both luminosity and higher surface temperature. Assuming that stars had fixed masses through their lifetimes, it was impossible for them to change their balance of temperature and brightness without major changes to their internal energy source. The implication was that stars following Eddington's model of stellar structure would sit in one location on the main sequence of the H–R diagram for most of their lives – a location determined at birth by the mass with which they formed.

This revolutionary new interpretation of the H–R diagram was greeted with resounding scepticism among Eddington's colleagues, not least because of the ongoing problem of stellar energy sources. Eddington himself had helped disprove the old gravitational contraction model, but the

The lifespan of stars

The duration of a star's lifetime can vary dramatically depending on its mass and composition. Heavyweight stars may have several times the mass of the Sun, but shine with many thousands of times its luminosity and therefore burn through their fuel much more quickly. While our Sun will spend about 10 billion years on the main sequence (fusing hydrogen into helium in its core) and hundreds of millions of years in the later stages of its evolution, a star of eight solar masses may exhaust its core supplies of hydrogen in just a few million years, with the other stages of its life cycle also dramatically shortened.

Mass is the most important factor affecting a star's lifespan. The higher temperatures and pressures in the core of massive stars allow the vastly more efficient CNO fusion cycle to become dominant (see page 75), while in less massive stars the more subdued proton–proton chain generates most of the energy. Composition also has a role to play: the CNO cycle can only take place if carbon is present to act as a catalyst, and since the Universe has only become enriched with carbon over time (see Chapter 42), the CNO cycle is less significant in earlier generations of stars.

1945
Gamow explains red giants as a late stage in the evolution of Sun-like stars

1956
Iosif Shklovsky shows planetary nebulae are red giants that have shed their atmospheres

1961
Chushiro Hayashi describes the paths of stellar evolution prior to the main sequence

STARS HAVE A LIFE CYCLE MUCH LIKE ANIMALS. THEY GET BORN, THEY GROW, THEY GO THROUGH A DEFINITE INTERNAL DEVELOPMENT, AND FINALLY THEY DIE.

Hans Bethe

favoured replacement was a hypothetical 'matter annihilation' model (see page 74) that would produce abundant energy and allow stars to shine for perhaps trillions of years, but also cause significant mass loss over a star's lifetime. Based on this assumption, evolution *down* the main sequence, with stars losing mass and fading as they aged, seemed to make sense.

It was only in the late 1930s, with Hans Bethe's breakthrough work on the proton–proton fusion chain (see page 74) that everything began to fall into place. In order to release the observed energy output of stars, nuclear fusion needed to take place at a much faster rate than the annihilation process, but it could still keep a star like the Sun shining steadily for billions of years. What was more, there would be relatively little mass loss between the beginning and end of the star's life.

EXPLAINING THE GIANTS

Eddington's argument that stars spent most of their life at one point on the main sequence was vindicated, but there were still major questions to be answered about how other types of star fitted into the story. As it happened, in the same year that Bethe published his thoughts on fusion, Estonian astronomer Ernst Öpik put forward a new view of stellar structure that also had important implications for evolution.

Eddington's assumption that material inside a star was constantly being stirred and mixed together had been widely accepted in the astronomical community, and implied that all of that material was ultimately available as fuel. Öpik, however, argued for a layered model in which the products of fusion remained in the core. This meant that the star's fuel supply was much more limited, and also ensured that the core grew slowly denser and hotter over time, as its hydrogen fuel was fused into helium. The churning convective core was surrounded by a far deeper hydrogen envelope in which energy was transported outwards primarily by radiation. All this material, making up the vast bulk of the star, was normally unavailable to act as fusion fuel, but this could change in older stars. Building on an idea first suggested by George Gamow, Öpik argued that proximity to an

1. Main-sequence fusion

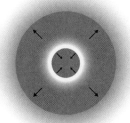

2. Hydrogen shell burning

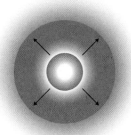

3. Core reignition

A star spends most of its lifespan fusing hydrogen into helium in its core [1]. When the core fuel supply is exhausted, hydrogen fusion moves into a surrounding shell [2]. Eventually, the contracting core grows dense and hot enough to support helium fusion [3].

increasingly hot core could heat the bottom of the radiative layer until it too became capable of sustaining fusion reactions. This would also cause the envelope to expand hugely in size.

Öpik's layered model would prove to be the key to explaining the patterns of evolution, but it was some years before it became widely accepted. George Gamow led attempts to model the structure of red giant stars and their position in the evolutionary sequence, but he was repeatedly misled by the belief that stars must be 'well mixed'. It was not until 1945 that he incorporated a layered approach into his model and showed that red giants are a late stage in the evolution of relatively normal stars, where hydrogen fusion in a shell around the core causes them to be both far brighter, and far larger than their main-sequence precursors. Gamow even realized that a red giant would ultimately shed its outer layers and expose its exhausted core as a hot but faint white dwarf (see page 124). These breakthroughs were just the tentative first step towards explaining the complicated story of post-main sequence evolution.

The condensed idea
The life cycle of a star is determined by its mass at birth

20 Nebulae and star clusters

Stars originate from huge collapsing clouds of interstellar gas, and their formation often lights up this gas to create spectacular nebulae. But while the association between tight star clusters and nebulae was recognized by the early 1800s, it took time to work out how one gives rise to the other.

The word *nebula* is Latin for 'cloud', and was used by stargazers as early as Ptolemy of Alexandria to describe the handful of fuzzy objects in the night sky that were not obviously composed of individual stars. It was only with the advent of the telescope, however, that astronomers began to discover many more of these objects. One of the first, and certainly the most famous, catalogues of nebulae was made by French comet-hunter Charles Messier in 1771, ostensibly to help avoid cases of mistaken identity when scanning the sky for comets.

Two decades later, William Herschel revisited the objects of Messier's catalogue with a more powerful telescope, and was able to distinguish between several different types of nebula. Some resolved themselves into groups or clusters of stars – or at least looked as if they might do so with an even more powerful instrument – whereas others seemed to be glowing clouds of gas, usually with stars or even open star clusters embedded within them.

TIMELINE

1771	1791–1811	1864
Messier draws up the first catalogue of non-stellar astronomical objects	Herschel recognizes gaseous 'bright fluid' nebulae, and links them to star formation	Huggins shows that the bright fluid nebulae are gaseous in nature

SITES OF STARBIRTH

Herschel called these clouds 'bright fluids' and they were the first conclusive evidence for matter in the space between the stars. Over the next two decades, he returned to study them occasionally, and in 1811 he outlined a theory that the bright fluids were the locations of star formation. Herschel believed that by looking at various nebulae, he was able to trace their condensation into individual stars and star clusters almost one step at a time. However, he made a significant mistake in assuming that stars were being formed individually and then pulled together by gravity to form clusters: in other words, the more tightly bound clusters were older than the looser ones.

The later 19th century saw significant advances in the study of both nebulae and open clusters. From 1864, William Huggins analysed the spectra of nebulae and showed that the light from Herschel's bright fluids consisted of a few narrow emission lines of specific colours, while that from other types of nebulae showed dark absorption lines against a broad continuum of different colours. This proved that the star-forming nebulae (now called emission nebulae) were largely gaseous in nature and suggested that many others, often spiral in shape, combined the light from vast numbers of stars (see page 146).

> WE MAY CONCEIVE THAT, PERHAPS IN PROGRESS OF TIME THESE NEBULÆ... MAY BE STILL FARTHER CONDENSED SO AS ACTUALLY TO BECOME STARS.
>
> William Herschel

In 1888, meanwhile, Danish–Irish astronomer J.L.E. Dreyer published the New General Catalogue (NGC), an extensive new listing of non-stellar objects in which he distinguished between two types of star cluster – tightly bound, spherical balls packed with thousands of stars, and looser groups with dozens or hundreds of members. The former were subsequently named globular clusters, but only the latter, dubbed open clusters, were found in association with emission nebulae.

1888	**1929**	**1947**
Dreyer distinguishes between open and globular star clusters	Hertzsprung develops methods for gauging the age of open clusters from the colours of their stars	Victor Ambartsumian identifies the first OB associations

Globular clusters

In addition to open clusters, Dreyer identified a second type of star cluster. These globular clusters have a far more concentrated structure and a completely different origin. They contain hundreds of thousands of stars, whose  elliptical orbits overlap to form a roughly spherical or elliptical shape. Individual stars are separated by light days or months, rather than years. Globular clusters are found close to the centre of galaxies, or orbiting in the halo regions above and below them (see page 137), and are almost entirely composed of low-mass dwarf stars with lifespans of many billions of years. Spectroscopic evidence suggests they lack the heavier elements found in more recently born stars, so they may have formed long before our Sun, in the early days of the Universe. In fact, the latest thinking links their origin to collisions between galaxies (see page 149).

The picture of star-forming regions grew yet more complex in the early 20th century, when the US astrophotography pioneer E.E. Barnard and his German colleague, Max Wolf, showed that they were often associated with opaque regions of light-absorbing dust ('dark nebulae'). In 1912, meanwhile, Vesto Slipher discovered yet another type of interstellar cloud in the Pleiades star cluster. This 'reflection nebula' shone by reflecting light from a nearby star.

DATING STAR CLUSTERS

While it seemed clear that emission nebulae were sites of starbirth, the actual sequence of events was frustratingly unclear. Breakthroughs in understanding of stellar life cycles and the evolution of clusters would begin to make sense of things, however. In 1929, for example, Ejnar Hertzsprung noted a significant difference in the properties of stars in the famous Pleiades, Praesepe and Hyades open clusters. The brightest Pleiades stars are all hot and blue, while Praesepe and especially the Hyades contain more bright orange and red stars. A few years later, it became clear that the colour differences were an indication of the relative ages of the clusters: the brightest and most massive stars shine hotter and bluer during their main sequence lifetimes, but age much faster, moving off the main sequence to become even brighter, though cooler, giants within just a few million years. Therefore, the older a cluster is, the more luminous red giants it contains.

The ability to put clusters in chronological order showed that Herschel's theory about clusters growing denser over time needed to be reversed. In fact, the densest clusters are the youngest and they grow successively looser over millions of years. In 1947, Armenian astronomer Victor Ambartsumian made a further breakthrough when he identified the first OB associations. These groups of fairly young, hot and bright stars are scattered across much wider areas of space, but show proper motions that can ultimately be traced back to the same point. Ambartsumian's discovery was the final confirmation that stars are born as tight open clusters within nebulae, before slowly scattering across space. Today, we know that this main dispersal mechanism involves close encounters between stars that end up being catapulted out of the cluster in different directions – sometimes at very high speeds.

The Ursa Major Moving Group

Sometimes, open clusters hang together for a surprisingly long time – for example, several dozen widely scattered stars, including five members of the famous Big Dipper, still share a common motion across the sky as the so-called Ursa Major Moving Group. This group, whose members all formed in the same nebula around 300 million years ago, was discovered by English astronomer and writer Richard A. Proctor in 1869.

By the mid-20th century, the locations of starbirth were beyond doubt, but it would take a revolution in observing technology for astronomers to really get to grips with the process involved (see page 85). Another key question was exactly what triggered the initial collapse of nebulae to create star clusters in the first place? Various mechanisms were put forward, from tidal forces raised by passing stars to the shockwaves from supernova explosions, but while chance events of this kind undoubtedly have a role to play, the principal mechanism would soon prove to be associated with the wider structure of our galaxy and others (see page 138).

The condensed idea
Gas clouds in space are the birthplace of new stars

21 Starbirth

By the mid-20th century, stars were understood to originate in dense star clusters, formed by collapsing gas clouds in emission nebulae. However, it took the arrival of space-age astronomy to reveal new detail within these emission nebulae and explain the specific processes involved in star formation.

The first clues to the exact mechanism of starbirth came in 1947, when astronomer Bart Bok highlighted the presence of relatively small, opaque clumps within star-forming nebulae. With diameters up to a light year across, Bok suggested these globules were cocoons, inside which individual star systems were forming.

For a long time this hypothesis remained unprovable, simply because 'Bok globules' are by their nature opaque. But as space-based astronomy began to develop in the 1970s, it was finally possible to address such problems. In particular, the InfraRed Astronomical Satellite (IRAS), an international collaboration launched in 1983, delivered a completely new view of the sky. IRAS was only operational for ten months, but in that time it mapped 96 per cent of the sky at four different infrared wavelengths, generating a dataset that kept astronomers busy for years.

LIGHT IN THE DARKNESS

Infrared radiation, with wavelengths longer and less energetic than visible light, is emitted by all objects in the Universe, and penetrates through opaque dust such as that within the Bok globules. In 1990, astronomers João Lin Yun and Dan Clemens announced that many globules coincided with infrared sources in the IRAS data, just as might be expected if they concealed young pre-main-sequence stars.

A few years later, in 1995, the Hubble Space Telescope snapped the famous 'Pillars of Creation' image. Zooming in on a star-forming region known as the Eagle Nebula (Messier 16) in unprecedented detail, it revealed towers of opaque gas and dust, from which strange trunks and tendrils emerged. Glowing halos around the pillars showed that they were evaporating under torrents of radiation from nearby massive stars. Jeff Hester and Paul Scowen, who made the image, interpreted the shape of the pillars as denser regions within the larger nebula that were better able to withstand the effects of radiation. Trunklike features emerge when a knot of material around a coalescing star (a Bok globule) remains intact even as its surroundings are beaten back.

Since that original image was taken, the Pillars and other star-forming regions have been imaged many times, both in visible and in infrared, and the same story seems to repeat time and time again. Intense radiation from an initial generation of massive and luminous newborn stars carves out hollows in the surrounding nebula. Pillars and tendrils emerge from its walls, marking sites where star formation is still ongoing. The effects of this radiation driving off the nebula's material – coupled with the shockwaves when these early stars explode as supernovae (see Chapter 30) – effectively limit the growth of their younger siblings in the cluster. According to a 2001 study, only about one-third of the gas in the original nebula ends up being incorporated into its stars, and the star-formation process lasts just a few million years at most before all the gas is lost. In a large majority of cases, the loss of so much mass causes the

> [T TAURI] STARS WERE BORN IN THOSE DARK CLOUDS... AND THERE HAS NOT BEEN ENOUGH TIME FOR THEM TO MOVE VERY FAR FROM THEIR BIRTHPLACES.
>
> George Herbig

1954	1961	1990	1995
Victor Ambartsumian suggests the Herbig–Haro objects form when T Tauri stars eject matter during their formation	Hayashi describes the details of pre-main sequence evolution in terms of tracks on the H–R diagram	Yun and Clemens link Bok globules to strong sources of infrared radiation, suggesting they have stars embedded within	The Hubble Space Telescope photographs the Pillars of Creation structures within the Eagle Nebula

Massive infant stars

In 1960, US astronomer George Herbig discovered a distinct class of unpredictable blue-white variables now called Herbig Ae/Be stars. They proved to be an early stage in the birth of stars heavier than the Sun (weighing from 2 to 8 solar masses). Like T Tauri stars, these monstrous stellar infants are surrounded by discs of material, some of which is still accreting onto them while much of the rest is being ejected into interstellar space. Research suggests that such young, high-mass stars do not follow the vertical Hayashi track on the H–R diagram at all. They are already highly luminous when they become visible and simply shrink over time, moving along the horizontal Henyey track and rapidly increasing their surface temperature in order to join the top end of the main sequence. Early stages in the evolution of the most massive stars of all (weighing tens of solar masses) are not so well understood, but it seems certain that they, too, move along the Henyey track at the beginning of their short lives.

nascent cluster to lose its gravitational integrity, suffering 'infant mortality' as its component stars and protostars drift apart. Only a minority survive to become mature open star clusters that contain between a hundred and a few thousand stars, and may hold together for tens of millions of years.

INFANT STARS

A single Bok globule may produce just one star, or a binary or multiple system if it condenses into two or more distinct cores. Computer modelling suggests that the initial collapse is quite rapid, with hot, dense protostars forming in tens of thousands of years. They remain embedded in a wider cloud of rotating matter, however, which gradually flattens out into a broad accretion disc. As the protostar's gravity grows stronger, it continues to pull in more material, but radiation from the increasingly hot core slows down the rate of infall and magnetic fields funnel material into jets that escape from above and below the disc. This combination of a strong stellar wind and magnetic fields is thought to transfer angular momentum to the disc, slowing the star's rotation and speeding up that of the surrounding material, resolving an old quandary about the origins of our own solar system (see page 17).

A star like the Sun may spend 10 million years or more as a protostar, emitting infrared radiation while growing steadily hotter and more energetic. Eventually, it starts to shine in visible light, whereupon it is said to have become a T Tauri star. These objects are large, reddish and more luminous than the stars they will become.

Since they get most of their energy from gravitational contraction rather than nuclear fusion reactions (see page 72), they vary unpredictably.

A typical T Tauri phase lasts 100 million years or more, during which time the star gradually contracts and its internal energy transport changes. In 1961, astrophysicist Chushiro Hayashi mapped out what these changes mean in terms of the Hertzsprung–Russell diagram. Pre-main-sequence stars become less luminous as they grow denser, initially allowing them to retain the same surface temperature. Stars with less than half the Sun's mass follow this 'Hayashi track' until their cores are dense enough for proton–proton fusion to begin, and they stabilize as red dwarfs (see page 88). Stars with up to twice the Sun's mass, however, change the direction of their evolution when their interiors become hot enough to develop a radiative zone (see page 70). They now retain the same luminosity, as they continue to shrink, leading to a rise in surface temperature (the so-called 'Henyey track'). In both cases, the beginning of proton–proton fusion marks the point where the star joins the main sequence, beginning the longest and most stable period of its life.

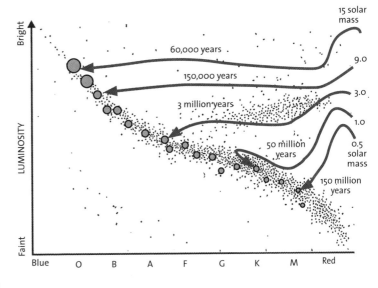

This H-R diagram shows the Hayashi and Henyey tracks of newborn stars approaching the main sequence.

The condensed idea
Infrared astronomy shows how stars are born

22 Dwarf stars

Stars with less mass than the Sun demonstrate their own unique and sometimes surprising properties, including surprisingly violent activity. At the very lowest masses, these red dwarfs shade into brown dwarfs, so-called failed stars whose existence has only been confirmed since the 1990s.

Technically speaking, nearly all stars are dwarfs, including our own Sun, and considerably more massive and luminous stars such as Sirius (see box, right). However, in common usage the term dwarf is used more specifically to describe small stars significantly less luminous than the Sun. Even this can be confusing, as white dwarfs, which are burnt-out stellar remnants (see page 124), are distinctly different objects from red dwarfs, which are just ordinary main-sequence stars of very low mass. They are both, in turn, different to brown dwarfs, which do not even meet the usual definition of a star.

The luminosities of stars vary far more widely than their masses, and just as stellar heavyweights can be a hundred thousand times brighter than the Sun, so the least massive stars can be a hundred thousand times fainter. A star with half the mass of the Sun (usually considered the upper limit for a red dwarf) shines with just 1/16th of its light, but a star with 0.2 solar masses has about 1/200th of the Sun's luminosity. This means that the vast majority of red dwarfs are very faint. For a long time the only known examples were those on our cosmic doorstep, such as Barnard's Star (see page 196) and Proxima Centauri, the closest star to the Sun. Despite being a mere

TIMELINE

1915	1962	1948
Robert Innes discovers Proxima Centauri, a faint red dwarf and the nearest star to the Sun	Kumar predicts the existence of abundant low-mass failed stars, later named brown dwarfs	Jacob Luyten discovers the nearby star BL Ceti, the first dwarf to show obvious flare star activity

4.25 light years away, this 0.12-solar-mass dwarf is 100 times fainter than the faintest naked-eye star, and was only discovered in 1915.

The abundance of red dwarfs in our galaxy only became clear with the launch of the first infrared space telescopes in the 1980s. The heat signatures of these dim stars are far more significant than their visible light output and maps of the infrared sky showed that red dwarfs vastly outnumber other stars, perhaps accounting for three quarters of all stars in the Milky Way.

DWARF STRUCTURE

An important difference between red dwarfs and more massive stars, and one that defines an upper mass limit for these stars, is the fact that they do not transport energy internally through radiation. Instead, their interiors are entirely convective, and the material they contain is constantly mixed and recycled. This mixing transports the helium products of nuclear fusion out of the core region and replaces it with fresh hydrogen, ensuring that all of the star's material is available for use as fuel for fusion. Coupled with the naturally slower rate that fusion proceeds thanks to the lower core temperature, this means that red dwarfs can theoretically sustain proton–proton fusion (and remain on the main sequence) for trillions of years – much longer than any other stars.

Defining dwarfs

According to Ejnar Hertzsprung's original definition, a dwarf is simply a star that obeys the widespread relationship between stellar temperature and luminosity, and therefore lies on the main sequence of the Hertzsprung–Russell diagram. The term dwarf was originally used to distinguish such stars from giants – the highly luminous stars of all colours found along the top of the H–R diagram, but the terminology has become confused over time.

What's more, at the upper left of the diagram, highly luminous blue dwarfs and giants are virtually indistinguishable based on colour and luminosity alone – they can only be separated if additional information confirms whether a star is still fusing hydrogen in its core. The use of the term white dwarf for burnt-out stellar remnants that lie nowhere near the main sequence (see page 124) only adds to the confusion.

1995

Rebolo *et al* discover the first confirmed brown dwarf star, Teide 1

2006

Michael Marks and Pavel Kroupa find a lower mass limit for stars of 0.083 solar masses, based on the faintest stars in a globular cluster

A red dwarf's core pumps out far less radiation than that of the Sun, meaning there is less outward pressure to support its outer layers. Therefore, these stars are a lot smaller and denser than their masses alone would suggest. Proxima Centauri is only 40 per cent bigger than Jupiter, and about 40 times denser than the Sun on average. This high density, together with a red dwarf's convective structure, can have unusual effects.

> **STARS WITH MASS BELOW A CERTAIN CRITICAL MASS WOULD KEEP ON CONTRACTING UNTIL THEY BECOME COMPLETELY DEGENERATE OBJECTS.**
>
> Shiv S. Kumar

The first evidence that dwarfs could show dramatic activity came from the Dutch-American astronomer Jacob Luyten, who discovered strange variations in the spectra of several nearby dwarfs in the 1940s. One in particular, the brighter star of a binary pair some 8.7 light years away in the constellation Cetus, was also prone to huge but short-lived increases in brightness. In a 1952 eruption, for instance, this star 'UV Ceti' increased in brightness by 75 times in a matter of seconds. By the 1970s, it was clear that the star's outbursts occurred not just in visible light but also in radio waves and high-energy X-rays, and were very similar to solar flares (see page 54) albeit on a much larger scale. Today, astronomers realize that many red dwarfs are also so-called flare stars. The density of these stars, and the convective churning of their interiors, generates far more powerful and concentrated magnetic fields than those seen in Sun-like stars. As a result, magnetic 'reconnection' events can release up to 10,000 times more energy than those that trigger flares on the Sun, with spectacular results.

BROWN DWARFS

According to most models of nuclear fusion, a star must have at least 0.08 times the mass of the Sun in order for temperatures and pressures in its core to sustain a proton–proton chain reaction. This is therefore the official cut-off for stars, but there are many objects below this mass that formed in the same way as stars, and may still pump out substantial amounts of infrared and visible radiation. Such 'failed stars', known as brown dwards, are kept hot by gravitational contraction and nuclear fusion of the heavy hydrogen isotope deuterium, which requires less demanding conditions. Their existence was theorized in the 1960s by astronomer Shiv Kumar (although the name was coined somewhat later).

During the 1980s, debatable objects with borderline properties were discovered, but in 1995, the first incontrovertible brown dwarf was found. Located by a Spanish team led by Rafael Rebolo, Teide 1 was a tiny object embedded in the distant Pleiades star cluster. Evidence of lithium in its spectrum was the giveaway clue to its identity, since even the lightest true stars are hot enough to destroy all trace of this element through nuclear fusion.

Hundreds more brown dwarfs have since been found, including many embedded in famous star-forming nebulae or on our cosmic doorstep, and often in orbit around other dwarf stars. Estimates of their masses suggest that the smallest brown dwarfs may actually be less massive than the biggest gas-giant planets the distinguishing factor between the two types of object is their mode of formation.

Brown dwarf weather

Just like stars, brown dwarfs can be classified by spectral type, according to their temperature and the absorption lines found in their atmosphere. The brightest brown dwarfs, like the faintest red dwarfs, have spectral class M (see page 65), but from here researchers have added the new classes L, T and Y. As these stars get successively cooler, increasingly complex molecules can persist in their atmospheres. Recent studies have shown variations in the infrared output of faint brown dwarfs that seem to be caused by enormous (planet-sized) cloud features moving in their atmospheres and temporarily blocking the escape of heat from within. The clouds are pushed around under the influence of extreme winds – as might be expected, weather on brown dwarfs is even more violent than that on gas giants such as Jupiter.

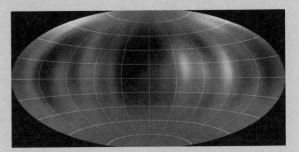

Weather map of the brown dwarf Luhman 16B

The condensed idea
The smallest stars are also the most abundant

23 Binary and multiple stars

Our Sun's status as a single star puts it in the minority. Today, we know that most stars in the Milky Way are members of binary or multiple star systems. The identical distance and age of the stars in such systems can reveal important facts about stellar evolution.

While the few thousand naked-eye stars visible from Earth seem to have a more or less random distribution across the sky, even an unaided stargazer can spot a couple of stars that appear to be exceptions. Close pairings in the sky are usually known as double stars, and perhaps the most famous example is Mizar and Alcor, the middle star in the tail of the Great Bear Ursa Major. Early astronomers thought little of this: if the distribution of stars was truly random, then a few such close pairs might be expected. However, when Italian astronomer Benedetto Castelli turned an early telescope on the star system in early 1617, he found something else. While Mizar appears as a single star to the naked eye, it is itself, in fact, a close pairing of two white stars, each of naked-eye brightness.

Such a close alignment of two bright stars just by chance is far less probable than the loosely associated Mizar and Alcor, but it was a long time before anyone properly considered the implications. The first person to suggest that the twin stars of Mizar were real neighbours in space was the English

TIMELINE

1617	1783	1804
Castelli discovers the first telescopic double star, Mizar in Ursa Major	Goodricke proposes an eclipse mechanism to explain the variable star Algol in Perseus	Herschel demonstrates that the two stars of Alula Australis are in orbit around each other

philosopher John Michell in 1767. Then, in 1802, William Herschel produced statistical evidence – based on a meticulous survey of the heavens – that close double stars were far too common to explain as chance alignments.

BINARY ORBITS

Herschel argued they must be 'binary' stars, physically held in orbit around each other by their mutual gravitational attraction. In 1804, he clinched the argument by showing that the two stars of Alula Australis (another close double in Ursa Major, which he had discovered 24 years previously) had altered their relative orientation, proving that they were orbiting each other. By 1826, the pair had been observed closely enough for French astronomer Félix Savary to analyse their orbit in detail. He showed that these two stars of roughly solar mass follow 60-year elliptical orbits with a separation that ranges between 12 and 39 astronomical units.

As telescopes improved in the 19th century, more binary stars, and even multiple systems containing more than two components, were discovered. It soon became clear that the distance between stars in such systems varies hugely: they may be separated by interplanetary distances of a few astronomical units, or interstellar distances of a light year or more.

Binaries and multiples also helped astronomers begin to understand the relationships within stars. For instance, since all the stars in a system are at the same distance from Earth, differences in their apparent magnitude correspond to differences in true luminosity. If we can establish the size of each star's orbit, then we can work out their relative masses, and since we can assume all of a system's stars formed at the same time, we can even start to see how properties such as mass affect a star's evolution over time.

> IT IS EASY TO PROVE … THAT TWO STARS MAY BE SO CONNECTED TOGETHER AS TO PERFORM CIRCLES, OR SIMILAR ELLIPSES, ROUND THEIR COMMON CENTRE OF GRAVITY.
>
> William Herschel

1889	1901	1903
Maury and Pickering identify Mizar A as the first spectroscopic binary star	Vogel deduces the physical properties of the two stars in Mizar A from spectroscopic data	Gustav Müller and Paul Kempf discover W Ursae Majoris, the first known contact binary system

Eclipsing binaries

In certain situations, binary stars too close to separate with a telescope can be detected through their effect on the light of the overall system. The first of these stars to be discovered, Algol (Beta Persei), has been known since ancient times as the 'winking demon'. It was described more or less accurately even before Herschel had clinched the case for physical binary stars.

In 1783, 18-year-old English amateur astronomer John Goodricke noted that Algol normally shines with steady brightness of magnitude 2.1, but dips abruptly to 3.4 for about 10 hours in a cycle that repeats every 2 days and 21 hours. He suggested this variability was best explained if Algol is orbited by a darker body, which passes across the star's face and partially blocks its light once in each orbit. The idea – the first proposed mechanism to explain a variable star of any kind – led to him being awarded the Royal Society's prestigious Copley Medal. It was not until the 1880s that Pickering and his Harvard team showed the fainter orbiting body is a star in its own right. Algol is today regarded as an eclipsing binary, a prototype of this important class of variable stars.

SPECTROSCOPIC BINARIES

Direct observation imposes limits on the types of multiple stars that can be discovered. No matter how powerful telescopes become, if the stars are too close together or too far from Earth, they will merge into a single point of light. In the late 19th century, however, a new method for discovering multiple stars was discovered.

Curiously, it was studies of Mizar that led the way once again. As part of his project to catalogue the spectral types and chemistry of stars (see page 63), Harvard astronomer William Pickering collected spectra from both elements of the twin star over 70 nights between 1887 and 1889, and tasked 21-year-old Antonia Maury with their analysis. Maury soon identified a strange feature in the spectrum of the brighter star Mizar A. The dark K line signifying calcium in its atmosphere appeared sharp and well-defined in some spectra, but broad and fuzzy in others, and on three photographic plates, it had split into two distinct lines.

Maury realized that this 'line doubling' effect was occurring every 52 days, and Pickering correctly identified the cause: Mizar A, in fact, consists of two stars in a tight orbit around each other. Both stars contribute to the overall spectrum, but the shifting K line reveals that their light outputs are continuously Doppler shifted as their motion relative to Earth changes (see page 62). At some points in their orbits, one star will be moving towards Earth, causing the wavelengths of its light to be shortened and the K line to shift towards the blue end of the spectrum, while the other star simultaneously moves away from Earth, its light stretching and reddening. At other times the situation is reversed, or the stars are moving sideways relative to Earth and so the Doppler effect disappears.

Mizar A became the first in a new class of spectroscopic binaries, which slowly revealed a truth that the vast majority of stars in our galaxy lie within binary or multiple systems. Just as importantly, Pickering realized that these stars offered a powerful new tool for astronomers: using the speed and orbital period, one can calculate the distance between the stars, and directly measure their masses using Newton's law of universal gravitation. It took some years to get the method right (Mizar A's orbit was eventually resolved by German astronomer Hermann Vogel in 1901), but the ability to directly measure such physical properties of distant stars, and indeed the mere existence of binary and multiple star systems, was to have a huge influence on 20th-century astronomy.

Contact binaries

In some binary systems, the components may be so close together that evolutionary changes in the size of one or both stars brings them into direct contact with each other. This happens when a star overflows the 'Roche lobe' that constrains its gravitational influence. In this scenario, the two stars form a W Ursae Majoris variable, an eclipsing binary whose light output is constantly varying. Substantial transfers of mass from one star to the other over millions of years can even alter the evolutionary pathway of one or both stars.

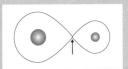

During their main-sequence lifetimes, both stars are within their Roche lobes.

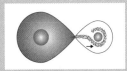

As the more massive star swells to a giant, material is transferred to its neighbour.

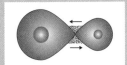

Eventually the smaller star may also swell into a giant. Material now flows in both directions.

The condensed idea
Single stars like our Sun are in the minority

24 Searching for exoplanets

Once astronomers began to accept that there was nothing particularly special about our Sun or our place in the Milky Way galaxy, it was hard to think of a reason why distant stars should not have planetary systems of their own. But proving it would take a long time, and it's only since the 1990s that so-called 'exoplanets' have been found in large numbers.

The search for planets orbiting other stars was long hampered by the limitations of technology. However, the discovery of Barnard's Star in 1916 brought with it the first hopes of discovering alien planets. This faint red dwarf was listed in star catalogues, but the astrophotographer E.E. Barnard was the first to recognize its unusually high proper motion against the background stars. Movement equivalent to the width of a full Moon every 180 years hinted that Barnard's Star was close to our own solar system, and parallax measurements (see page 58) soon confirmed that it was just six light years from Earth, the fourth-closest star to the Sun.

A FALSE START

Dutch astronomer Peter van de Kamp soon realized that the fast motion of Barnard's Star would make any wobbles in its path, due to the gravitational pull of large planets, particularly noticeable. For over three decades beginning in 1937, he regularly tracked the star's precise position,

TIMELINE

1969	1992	1995
Van de Kamp publishes mistaken evidence for planets orbiting Barnard's Star	Wolszczan and Frail discover the first known pulsar planets orbiting PSR B1257+12	Mayor and Queloz announce the discovery of 51 Pegasi b, the first exoplanet around a normal star

and eventually, in 1969, he published evidence for two Jupiter-class planets. But his observations proved hard for others to replicate, and by the 1980s most concluded Van de Kamp had been mistaken, perhaps due to faults in his equipment. The Barnard's Star affair left many astronomers sore and a more sceptical opinion took hold. Most assumed that planets around other stars were very rare for some reason. Happily, it wasn't long before this stance was itself undermined by a new and more sensitive planet-finding method.

SUCCESS AT LAST

The idea of detecting planets around other stars by measuring changes in their *radial* velocity (motion towards or away from Earth) was proposed as early as 1952 by Otto Struve. This Ukrainian–American astronomer suggested that just as 'spectroscopic' binaries reveal their true nature through back-and-forth Doppler shifts of spectral lines as their components move towards and away from Earth (see page 94), so the influence of a planet on its star should show up if a sufficiently sensitive spectrograph was employed.

The problem, however (as Van de Kamp had discovered) is that a planet wields very little influence over its star. Depending on their relative masses and the size of the planet's orbit, the biggest disruption one could hope

Pulsar planets

The very first planets to be discovered around another star were actually found a few years before 51 Pegasi b. However, they gained rather less attention because the conditions in which they were found are inimical to life. In 1992, astronomers Aleksander Wolszczan and Dale Frail announced the discovery of two planets orbiting a pulsar designated PSR B1257+12, some 23,000 light years away in the constellation of Virgo (see page 127). A third planet followed in 1994. Several planetary systems have now been found through careful analysis of tiny changes in the otherwise-precise flashes of radiation, which occur as a pulsar is pulled in different directions by its orbiting worlds. Such planets are unlikely to have survived the supernova explosion in which the pulsar formed. Instead, they are thought to be born in a secondary phase of planet formation, out of debris from a destroyed companion star.

1999

The first exoplanet is discovered using the transit method

2009

NASA launches its Kepler planet-hunting mission, leading to the identification of thousands of new exoplanets for would be an oscillation on the order of a few metres per second, in an average velocity typically measured in *kilometres* per second. Detecting such tiny variations would mean splitting the star's light into a very broad 'high-dispersal' spectrum, which was beyond the technology of the time. However, advances in the 1980s produced the first 'echelle spectrographs', suitable for analysing the faint light of stars. These instruments use a pair of diffraction gratings to create a broad spectrum, in conjunction with optical fibres to feed light from individual stars to the gratings.

The ELODIE instrument, operated by Michel Mayor and Didier Queloz at the Observatoire de Haute-Provence from 1993, was designed specifically to look for exoplanets, and soon proved its worth. In 1995, Mayor and Queloz were able to announce the discovery of a planet with at least half the mass of Jupiter in orbit around the relatively nearby star 51 Pegasi. This was the first of several such discoveries from ELODIE and its southern-hemisphere counterparts.

THE MISSION OF NASA'S KEPLER TELESCOPE IS TO LIFT THE SCALES FROM OUR EYES AND REVEAL TO US JUST HOW TYPICAL OUR HOME WORLD IS.

Seth Shostak

TRANSITS AND OTHER METHODS

A few years after those early discoveries, an even more effective technique had its first success. The transit method involves measuring the tiny drop in a star's overall light output when a planet passes directly in front of it. Because the star's size is relatively easy to work out from its spectral features (see page 65), the relative dip in its light output reveals the size of the transiting planet. Clearly, transits only happen in rare cases when a planet's orbit is directly aligned with Earth, but given the sensitivity of modern light-measuring photometers, it's currently the most practical way of identifying low-mass, Earth-sized exoplanets.

The first transiting exoplanet to be discovered, in 1999, orbits an obscure Sun-like star catalogued as HD 209458, 150 light years away in Pegasus. Astronomers already knew this star had a planet in a tight orbit thanks to radial velocity measurements, but the transit confirmed that it had a radius roughly 1.4 times that of Jupiter. Since that first discovery from the Keck Observatory on Hawaii, satellite-based, transit-spotting telescopes have been the most successful method of planet-hunting. The first, a French mission called COROT operated between 2006–2012, while the NASA instrument, Kepler (see box), came later. A location in orbit allows a telescope to continuously monitor the brightness of a whole field of stars for long periods without interruption, making it easier to detect planets in longer orbits.

PLANETARY PROPERTIES

Different types of planet-hunting technique reveal different physical properties of exoplanets. The radial velocity method, for example, puts a minimum mass on the planet causing the star to wobble, but unless the inclination of the planet's orbit is known, it cannot find a more precise value for the mass. The transit method, in contrast, can reveal the diameter of a planet, but not its mass. In practice, observing a planet both ways reveals the most information. If radial velocity data can be obtained, then he mere fact that a planet transits its star constrains its orbital inclination and possible mass, which together with a diameter measurement can confirm the planet's density and allow astronomers to work out its likely composition.

Kepler

Launched in 2009, NASA's Kepler satellite is a dedicated planet-hunting spacecraft that has transformed our knowledge of exoplanets. Its single instrument is a 0.95-metre (37-inch) reflecting telescope attached to a photometer camera that measures tiny variations in starlight to detect planetary transits. During its primary mission, four 'reaction wheels' were used to keep Kepler's gaze precisely fixed on a single field of view – a swathe of the Milky Way mostly in the constellation of Cygnus. Following the failure of two of these wheels and loss of precise tracking, engineers found an ingenious way to keep the telescope oriented in space using the pressure of radiation from the Sun. This enabled shorter but still useful periods of tracking stars. So far, Kepler has discovered more than a thousand exoplanets, with several thousand more awaiting confirmation.

The condensed idea
Looking for planets around other stars needs ingenious techniques and sensitive instruments

25 Other solar systems

Before the discovery of the first exoplanets, astronomers had assumed that alien solar systems would follow a pattern similar to our own. Recent discoveries, however, have revealed a whole range of unexpected new planet types and orbits, suggesting that planetary systems evolve significantly throughout their history.

From the moment in 1995 that Mayor and Queloz announced their discovery of 51 Pegasi b – the first confirmed exoplanet around a Sun-like star (see page 98) – planetary scientists found themselves faced with a puzzle. The new planet orbited its star in just 4.23 days, seven times closer than even Mercury is to the Sun. What was more, the planet's mass was at least half the mass of Jupiter (perhaps significantly more). What was a probable gas-giant planet doing so close to its star?

As new worlds began to emerge at an ever-increasing rate, it soon became clear that 51 Pegasi b was no one-off. In fact, a substantial fraction of all early discoveries turned out to be so-called 'hot Jupiters' – giant planets in tight orbits around their stars. This was partly a consequence of the radial velocity method used to make these initial discoveries: only substantial planets have enough mass to affect the Doppler shift of their star's light, and the repeating shifts due to planets in short-period orbits will be the easiest to pick out. The transit method, too, is biased towards finding planets close to their stars, not only because their transit events take place more frequently, but also because the chances of a transit-causing alignment are dramatically higher for planets with smaller orbits.

TIMELINE

1995	2005	2007
Mayor and Queloz discover the first 'hot Jupiter', 51 Pegasi b	Eugenio Rivera *et al.* discover Gliese 867 d, the first super-Earth around a main-sequence star	Snellen *et al.* deduce the presence of high-speed winds in the atmosphere of HD 209548 b

But this does little to alter the fact that, according to otherwise-successful models of planetary formation (see page 18), gas-giant-type planets should not be able to form so close to a star.

PLANETS OUT OF PLACE

A potential solution to the hot Jupiter problem arises from theories of planetary migration, the idea that planets shift their positions significantly over long periods of time. Given the right initial conditions, it's not too difficult to model a scenario in which a giant planet starts life beyond the snow line of its solar system where gas and ice are abundant, but then spirals inward thanks to tidal interactions with gas in the protoplanetary nebula. One sinister possibility is that a giant planet on such a slow inward track would disrupt the orbits of any worlds that had formed closer to its star – precisely the kind of small, rocky worlds that might be hospitable to alien life.

Hot Jupiters have so far been discovered with a wide range of masses, varying from a little less than Jupiter itself, up to around ten times heavier, about the same as the smallest 'brown dwarf' stars (see page 90). At the less massive end of this range, heat from the nearby star can make the planet's atmosphere balloon in size against the comparatively weak gravity, creating a low-density 'puffy planet'. This theoretically predicted effect has been confirmed by subsequent observations of transiting exoplanets whose diameter can be calculated directly.

However, some other, more massive planets with higher gravity seem to be larger and hotter than theory predicts. In 2013, Derek Buzasi of Florida Gulf Coast University identified a potential link between these larger-than-expected planets and the magnetic activity of their parent stars, suggesting that magnetism may play a substantial role in heating them up.

> **THERE SEEMS TO BE NO COMPELLING REASON WHY THE HYPOTHETICAL STELLAR PLANETS SHOULD NOT... BE MUCH CLOSER TO THEIR PARENT STARS.**
>
> Otto Struve, 1952

2009

Launch of Kepler transforms the kinds of exoplanets that can be discovered

2012

Nikku Madhusudhan *et al* identify 55 Cancri e as a possible carbon planet

Measuring planetary atmospheres

So far, it's only possible to image the direct light of exoplanets in very rare cases. However, observations of transiting exoplanets can occasionally yield data about their atmospheres. As a planet passes in front of its star, the gases in its atmosphere absorb certain wavelengths of light, altering the pattern and intensity of the star's own absorption spectrum (see page 60). In 2001, this technique was used to identify sodium in the atmosphere of HD 209548 b, a hot Jupiter some 154 light years away in Pegasus. Further studies of this intriguing planet revealed an envelope rich in hydrogen, carbon and oxygen that extends to more than twice its own radius. This is a sign that the planet is losing its atmosphere in the blast of heat from its parent star, which raises its temperature to around 1,000°C (1,800°F). By measuring the Doppler shift of carbon monoxide absorption lines in the planet's atmosphere, a team led by Ignas Snellen of Leiden University in the Netherlands not only measured the planet's precise speed in orbit, but also detected the presence of high-speed winds in its atmosphere, blowing at between 5,000 and 10,000 kilometres per hour (3,000–6,000 mph).

AN EXTRASOLAR ZOO

Hot Jupiters were the first of several new classes of planet that have emerged from observational data and computer modelling since the 1990s. These include:

• Hot Neptunes. As their name suggests, these planets are Neptune-mass giants in close orbits around their stars. Surprisingly, some models of planet formation suggest that giants of this class could potentially form at an Earth-like distance from their parent stars, with migration not necessarily involved.

• Chthonian planets. Several systems have been discovered in which radiation and stellar winds are stripping away the outer layers of a hot Jupiter, forming a comet-like tail. Chthonian planets are the hypothetical end result of this process. The merciless solar wind would leave just the exposed rocky core of a once-giant planet, reduced to an Earth-like mass.

• Super-Earths. These planets have a mass between around 5 and 10 Earth masses. Observations suggest that super-Earths have a variety of densities and therefore a range of compositions. Some may simply be oversized rocky planets, while others may be 'gas dwarfs'. Proximity to the central star determines surface conditions, which could potentially range from semi-molten lava seas to deep-frozen ice. Ocean planets are a particularly intriguing subgroup with a high proportion of water, thought to form when an initially icy world migrates closer to its star.

COMPOSITION

	Iron	Silicate (Earth-like)	Carbon	Water	Carbon monoxide	Pure hydrogen
Earth-mass planet	○	○ Earth equivalent	○	○	○	◯
Super-Earth	○	○	○	○	○	◯ 20,000 km (12,400 miles)

MASS

This table shows how the size of more-or-less Earth-like exoplanets varies depending on both mass and composition.

• **Carbon and iron planets.** Depending on conditions in the initial protoplanetary nebula, terrestrial planets may end up with much larger quantities of carbon or iron, rather than the silicate rock that dominates on Earth. Iron-dominated worlds may also be created when a planet is bombarded by major impacts that strip away the lighter elements in its mantle. In our own solar system, something like this is thought to have happened to Mercury.

So far, the study of these burgeoning objects is in its infancy, but it's already proving possible to identify a surprising range of physical characteristics in planets we cannot yet observe directly. Planning is already underway for a next generation of giant telescopes that will be able to resolve and study individual exoplanets, revealing even more about these intriguing and varied worlds.

The condensed idea
Our solar system's configuration is just one of many possibilities

26 Goldilocks zones

The search for truly Earth-like planets, potentially capable of sustaining life using carbon-based biochemistry, is one of the biggest challenges in modern astronomy. Yet understanding exactly what creates the 'habitable zone' around a particular star has proved to be a surprisingly complex task.

The idea of a star's radiation characteristics affecting the habitability of planets around it was put into writing in 1953 by two separate researchers: the Germany-born physician Hubertus Strughold and US astronomer Halton Arp. The fact that planetary conditions are hot close to the Sun and get colder further out in the solar system had been taken as read for centuries, but Strughold was the first to define 'zones' in which life was more or less likely, while Arp calculated the range of conditions in which liquid water could persist on a planetary surface. In 1959, Su-Shu Huang brought these concepts together into the idea of a 'habitable zone', based on what was then known about the origins of life and the conditions it required.

DEFINING THE GOLDILOCKS ZONE

Since then, the habitable zone – popularized since the 1970s as the 'Goldilocks zone' – has become a widely understood way of thinking about the prospects for life around other stars. Announcements of new exoplanets often focus on how Earth-like they are, with their position in the zone as a key factor.

According to the children's story, the Goldilocks zone should be where things are neither too hot nor too cold, but 'just right'. This might seem

TIMELINE

1953	1959	1979
Strughold and Arp independently study factors that affect the temperature and habitability of planets around other stars	Su-Shu Huang combines Strughold and Arp's ideas into the concept of a habitable zone around every star	The discoveries of tidal heating and ocean moons open up possibilities for life outside of the habitable zone

easy enough to calculate: for any given star the region must lie in between the points where its radiation heats a planet's surface enough to vaporize water (boiling point) and where it is insufficient to melt ice (melting point). Unfortunately, it's not that simple: in order to hold onto liquid water, a planet needs a reasonably substantial atmospheric pressure. Without it, liquid water simply boils away whatever the temperature. The lower the pressure, the lower the boiling point of water, as generations of disappointed mountaineers have found when trying to make a decent cup of tea.

The ability to retain an atmosphere is itself a function of a planet's mass and its position relative to its star. High gravity and/or cold conditions make it easier to prevent constantly shifting gases from drifting away into space. Any atmosphere has an insulating effect that helps even out temperatures between the planet's day and night sides, by preventing daytime heat from radiating away immediately after sunset. However, the precise chemical composition of an atmosphere also has a significant effect. Greenhouse gases, such as carbon dioxide, methane and water vapour, absorb much more of the escaping

Carbon chauvinism?

Most thinking about habitable zones implicitly accepts that life elsewhere in the Universe will be more or less similar to life on Earth. As early as 1973, however, planetary scientist Carl Sagan warned that such 'carbon chauvinism' could be misleading. In reality, there are good reasons to assume some essentials of life would remain the same throughout the galaxy. By most definitions, even the simplest forms of life involve some kind of information-carrying molecule analogous to DNA and capable of being inherited when an organism replicates itself. Carbon can reasonably be seen as the most likely basis for such a molecule because this abundant element's structure lets it form a unique variety of complex chemical bonds (other elements such as silicon and germanium form bonds in a similar way, but are less chemically reactive). The key role of water, meanwhile, is based on the simple need for a fluid medium in which chemical compounds can move around and undergo the reactions necessary to build complex molecules in the first place. Other liquids such as ammonia could theoretically play this role, but so far as we know water is both the most abundant potential medium, and also the one that remains in a liquid state over the broadest range of temperatures.

1987	1993	2011	2014
Marochnik and Mukhin formulate the idea of a galactic habitable zone, looking at regions in our galaxy that might support life	Kasting *et al.* introduce a new definition of the Goldilocks Zone that tends to shift it outwards from the central star	Astronomers discover Kepler-22b, the first known exoplanet orbiting within the habitable zone	Discovery of Kepler-186f, the first Earth-sized planet in the habitable zone

heat and keep a planet's surface relatively warmer. This effect is seen most starkly on Venus, where a dense carbon-dioxide atmosphere heats the surface by hundreds of degrees above what it would otherwise be.

Since it's still impossible for us to directly analyse the atmospheres of most exoplanets, exobiologists use standardized models to anticipate their warming effects. In 1993, geoscientist James Kasting and others modelled the Goldilocks Zone as the region between an inner edge, where water would always be lost from a planet with Earth-like gravity regardless of atmospheric composition, and an outer edge where water would be just above freezing point in a 'maximum greenhouse' (carbon-dioxide-dominated) atmosphere. Kasting's estimates put our own solar system's habitable zone at between 0.95 and 1.67 AU from the Sun, suggesting that Earth is skirting perilously close to the inner edge. In 2013 a new model pushed the habitable zone even further out, to between 0.99 and 1.70 AU.

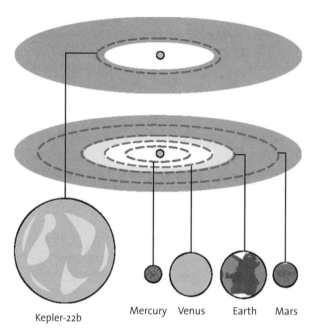

Kepler-22b

Mercury Venus Earth Mars

A comparison between Kepler-22b, the first exoplanet found in the Goldilocks zone around a Sunlike star, and the orbits of our own inner solar system.

SHIFTING THE GOALPOSTS

But even as some scientists have concentrated on refining the position of this 'traditional' Goldilocks zone, so new discoveries have added to the difficulty of defining where the habitable zone really lies, and shown that it's certainly not the last word in the search for life. The revelation of abundant extremophile organisms on Earth (see page 49) has shown that life can flourish in a much wider range of environments than previously thought, while the discovery of tidal heating effects and subterranean oceans among the icy moons of the outer solar system has broadened the parameters for potentially life-supporting worlds well beyond more conservative estimates of the Goldilocks zone.

Others, meanwhile, have extended the concept of habitability still further, narrowing the options for exoplanets with the potential for life. One possible consideration is a star's location in the wider galaxy. According to this 'galactic habitable zone' idea, stars in the crowded heart of a galaxy are more likely to be blasted by the sterilizing rays of supernova explosions, while stars near the outer edge will form without the dust required to make terrestrial planets in the first place. Some astronomers, however, doubt whether a star's position is quite so useful. Nikos Pranzos of the Paris Institute for Astrophysics has argued that there are simply too many variables involved, not least of which is the fact that a star's orbit within a galaxy may change considerably through its lifetime.

One other consideration is not spatial but temporal: based on the example of our own planet, evolving advanced life seems to take *time*. In Earth's case, the most primitive single-celled life forms got a foothold within about a billion years of Earth's formation, but it took a further 3 billion years for multicellular life to explode. This would seem to limit the potential for life to stars with multi-billion-year lifespans – in other words, those with not much more mass than our Sun. Some have argued that, because it would also have taken time for our galaxy to develop the heavy elements needed to form Earth-like planets, our generation of worlds may be the first with the potential for advanced life.

> **THE FIRST SIGNS OF OTHER LIFE IN THE GALAXY MAY WELL COME FROM PLANETS ORBITING AN M DWARF.**
>
> Elisa Quintana, SETI institute

The condensed idea
Conditions for life may exist on many exoplanets

27 Red giants

Amongst the largest stars in the Universe, red giants are the most spectacular stage in the evolution of stars like our Sun, and play a key role in creating heavy elements. Once thought to be infant stars, their true nature was only recognized once misconceptions about stellar structure had been laid to rest.

The term 'red giant' arose from Ejnar Hertzsprung's 1905 division of stars, by luminosity, into dwarfs and giants. Both he and Henry Norris Russell realized that the high luminosity and low surface temperature of these stars indicated enormous size. However, it was also clear that despite their prominence in Earth's skies, such luminous stars are extremely rare compared to their fainter dwarf brethren.

EXPLAINING MONSTERS

Red giants are so large that if such a star were to replace the Sun in our solar system, it would engulf the orbits of several planets including Earth. As early as 1919, Arthur Eddington predicted the size of the well known red giant Betelgeuse in the constellation of Orion. The following year Albert Michelson and Francis Pease targeted Betelgeuse using the Hooker Telescope at California's Mount Wilson Observatory, which was then the largest in the world, to confirm Eddington's estimate. Yet curiously, the evidence seemed to suggest that red giants did not *weigh* significantly more than normal dwarf stars. Clearly, there must be some fundamental difference between the processes creating energy in dwarfs and giants, but what could that be?

TIMELINE

1920	1938	1945
Michelson and Pease confirm the huge diameter of the star Betelgeuse in Orion	Öpik introduces the idea of fusion shells whose development triggers changes in a star's size and luminosity	George Gamow models red giants as a late stage in the evolution of Sun-like stars

The solution arose from Ernst Öpik's daring 1938 suggestion that the stars are *not* homogeneous (see page 78). Against the prevailing theories of the time that star interiors are well mixed, he proposed that energy production takes place in a discrete core region where the products of hydrogen fusion accumulate over time. By applying Eddington's idea of a balance between the outward force of radiation and the inward pressure of gravity, Öpik showed how the core would grow denser and hotter as it used up its supply of hydrogen. Eventually, despite the core's fuel becoming exhausted, the effect of its heat on the surroundings creates conditions suitable for fusion in a shell of material around it.

Due to the higher temperatures involved, this 'hydrogen shell burning' takes place at a much faster rate than core fusion, boosting the star's luminosity and causing the envelope region above the shell to expand enormously and create a red giant. Because shell burning squanders fuel rapidly, Öpik saw that it would be a relatively brief phase in a star's life cycle, explaining why red giants are so much rarer in our galaxy than dwarf stars.

> THE TIME SPENT BY A STAR IN... ITS EVOLUTION AS A RED GIANT MUST BE CONSIDERABLY SHORTER THAN THE PERIOD IT WILL SPEND IN THE MAIN SEQUENCE.
>
> George Gamow

BEYOND THE HYDROGEN SHELL

By the early 1950s, the ideas of hydrogen fusion as the main source of stellar energy, and shell burning as the driver for red giant evolution, were well established. The next obvious question was whether other fusion reactions could also play a role. Helium was particularly interesting, since it is produced in abundance by the earlier stages of hydrogen fusion. Various astrophysicists and nuclear scientists began to zero in on a particular chain of helium fusion reactions as a possible way that stars might keep shining and also generate some of the Universe's most abundant heavier elements. The solution came in the form of the triple-alpha process (see box, page 110). This is a fusion reaction between

1952

Hoyle and Fowler discover the triple-alpha helium fusion process

1956

Shklovsky shows that red giants shed their atmospheres in planetary nebulae, exposing their cores as white dwarf stars

1962

Schwarzschild and Härm identify the helium flash – a sudden change in red giant structure triggered by the onset of helium burning helium nuclei that ignites when the slowly collapsing core of a red giant star reaches a critical density and temperature.

Once helium-burning becomes possible, it rapidly spreads through the core in an event called the 'helium flash'. This reignition in the core has a significant effect on the star's internal structure. The restored radiation pressure from the core causes the hydrogen-burning shell to expand and become less dense, with fusion throttling back. As a result, the star as a whole contracts and becomes slightly less luminous. The core's helium supplies are exhausted quite rapidly, after which helium fusion moves out into a shell of its own underneath the hydrogen-burning shell, and the star once again brightens and expands. For the vast majority of stars, the end is rapidly approaching – the core, now rich in carbon and oxygen, continues to contract but will never reach the extreme temperatures necessary for another stay of execution.

Late in their evolution, many red giants develop pulsations in their outer layers, growing and shrinking due to instabilities in their internal structure (see page 112). These oscillations can be accompanied by substantial changes in luminosity that may be either a direct change in the star's energy output, or the result of opaque layers of gas and carbon-rich dust being cast out of the upper atmosphere and obscuring the light of the photosphere beneath.

The triple-alpha process

The mechanism responsible for fusing helium into carbon in evolved stellar cores is known as the triple-alpha process because a normal helium nucleus (consisting of two protons and two neutrons) is equivalent to the alpha particles emitted by some radioactive substances. The first stage of the process involves two helium nuclei uniting to form a nucleus of beryllium-8. This isotope of beryllium is highly unstable and normally disintegrates back into two helium nuclei almost immediately, but when conditions in the core pass a certain threshold, helium nuclei can make beryllium faster than it can disintegrate. As beryllium starts to build up in the core, the second stage of the process becomes possible – fusion with a further helium nucleus to create carbon. According to early 1950s models of nuclear interactions, this process would be highly unlikely to happen even when beryllium and helium nuclei are forced together, but British astrophysicist Fred Hoyle famously realized that it must occur if stars are to form carbon. He therefore predicted the existence of a 'resonance' between the energies of the three nuclei involved that would make fusion more likely. Despite the scepticism of the nuclear physics establishment, just such a resonance was subsequently discovered by William Alfred Fowler's team at the California Institute of Technology in 1952.

THE FINAL END

Soviet astronomer Iosif Shklovsky explained the fate of red giants in 1956. He found an evolutionary 'missing link' in the form of planetary nebulae. These beautiful hourglass- and ring-shaped bubbles of interstellar gas, which are illuminated by a hot white star at their centre, appear to be expanding at tremendous speed. Shklovsky realized that these features would make planetary nebulae very short-lived objects in astronomical terms (lasting perhaps just a few thousand years), and concluded that they must be an intermediate phase between two other, more widespread objects. The hot white central star seems to be hotter version of a 'white dwarf' (see page 124) – the eventual fate of all planetary nebulae. The gaseous shells, meanwhile, showed strong resemblance to the atmospheres of red giants – could this be their origin?

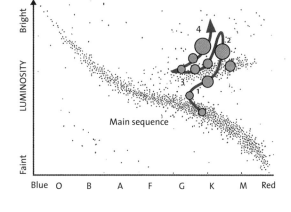

Later research conclusively backed up this daring synthesis. In 1966, George Ogden Abell and Peter Goldreich showed exactly how a red giant's atmosphere could escape to become a planetary nebula, while between the 1950s and 1970s Martin Schwarzschild and Richard Härm at Princeton used computers to model the entire story of red giants with increasing complexity. More recently, images from the Hubble Space Telescope and other modern observatories have revealed ever more detail about the dying stages of stars like the Sun.

As it exhausts the hydrogen in its core, a Sunlike star moves off the main sequence of the Hertzsprung-Russell diagram [1] brightening and swelling to become a red giant [2]. Core helium burning sees it move onto the horizontal branch [3], but when helium burning also moves out into a shell, the star swells again and moves onto the so-called asymptotic giant branch [4].

The condensed idea
Red giants are aged and evolved Sun-like stars

28 Pulsating stars

While the vast majority of stars shine with more or less steady luminosity for most of their lives, some vary in brightness significantly on relatively short timescales. Some of these are eclipsing binary stars, but changes in luminosity can also be the result of a single star undergoing dramatic pulsations.

The first pulsating variable star to be discovered is still the most famous. Catalogued as Omicron Ceti, the red star in the neck of the sea monster constellation Cetus undergoes dramatic variations in brightness that take it from an easy naked-eye object to one only visible through telescopes, in a cycle lasting about 11 months. First noted by David Fabricius in 1596, Johannes Hevelius soon named it 'Mira' (literally 'the astonishing one').

Variables were only discovered in greater numbers from the late 18th century onwards, and soon revealed their bewildering variety. While some were obviously red stars like Mira with long-period pulsations, others such as Delta Cephei in the constellation Cepheus, varied less dramatically and in periods of just a few days.

While Mira's variations could be somewhat erratic, the so-called Cepheids were soon discovered to repeat their cycle with metronomic precision. Development of photometric techniques for high-precision measurement of stellar magnitudes in the 20th century revealed an even greater array of changes, including stars that alter their brightness by fractions of a

TIMELINE

1596	1784	1879
Fabricius notices the changing brightness of Mira, the first variable star to be discovered	John Goodricke identifies the variability of Delta Cephei	Ritter suggests that pulsating stars are a result of internal changes rather than interactions with other stars

magnitude in periods of minutes, and more complex patterns made up of multiple overlaid pulsations.

INTERNAL CHANGES

German engineer August Ritter was the first to suggest that brightness variations in these stars were due to innate changes in their radius and brightness in 1879. At the time, most astronomers believed that variability was solely a product of interactions within binary star systems (see page 94), and so his ideas were largely ignored. In 1908, however, Henrietta Swan Leavitt (one of E.C. Pickering's 'Harvard Computers' – see page 63) made an important discovery. Among thousands of Cepheid variables photographed in the Small Magellanic Cloud (an isolated star cloud now known to be a satellite galaxy of the Milky Way), there seemed to be a clear relationship: the brighter a star's average appearance, the longer its cycle of variation. Assuming that the cloud was a physical object at a relatively large distance from Earth (so that all its stars are effectively at the same distance, and differences in apparent magnitude represent differences in intrinsic luminosity), Leavitt was able to conclude that there was a genuine relationship between period and luminosity at work.

In 1912, Leavitt published more detailed evidence for the relationship. Her discovery overturned long-held ideas about variable stars, since there was no plausible explanation for why an eclipsing binary or similar system would obey this period-luminosity rule. It would also later play a key role in developing ideas about the large-scale Universe (see page 146).

Despite the evidence assembled in 1914 by Harlow Shapley that Cepheids were driven by some kind of pulsation mechanism, a detailed explanation remained elusive. Then, in the 1920s, Arthur Eddington used his new model of

> **IT IS WORTHY OF NOTICE THAT... THE BRIGHTER VARIABLES HAVE THE LONGER PERIODS.**
> Henrietta Swan Leavitt

1908	1926	1953
Leavitt identifies the period-luminosity relation in Cepheids, supporting the idea that their changes are internal	Eddington shows that stellar pulsations are probably due to internal opacity changes	Zhevakin shows that hydrogen ionization can cause the opacity change in Eddington's model

Other types of variable

Not all intrinsically variable stars can be explained through Eddington's pulsation mechanism. Many young stars such as T Tauri stars (see page 86) fluctuate in brightness because their interiors have not yet reached equilibrium, and they may still be gaining or shedding considerable amounts of matter. Massive, highly luminous supergiants, meanwhile, can vary their light output because the sheer amount of radiation pressure they generate renders them unstable, often leading them to shed their outer layers into surrounding space (see page 117).

Other variables require a completely different explanation. These include R Coronae Borealis stars – giants that occasionally expel clouds of opaque dust, blocking much of their light from view for years. Astronomers have also only recently begun to recognize a wide variety of rotating variables. These are stars whose brightness varies as they spin on their axes, due either to huge dark starspots in their atmospheres, the effects of powerful magnetic fields or even – in the case of the fastest spinners and stars in close binary systems – distortions to their overall shape.

stellar interiors, to argue that pulsations must be regulated by a natural 'valve', which limits the radiation escaping from the star's surface.

Moreover, he showed how this situation could arise if a particular layer of the star's interior grew more opaque. The increasing density of a layer due to compression would tend to slow the escape of radiation, but the resulting increase in pressure from beneath would eventually push the layer outwards, whereupon it would become more transparent and allow the excess energy to escape. In this way, the process becomes a repeating cycle.

There was just one major problem with Eddington's theory – evidence suggests that increasing pressure in most regions of a star actually *reduces* its opacity (an effect known as Kramer's Law). It wasn't until the 1950s that Sergei Zhevakin found a mechanism to explain Cepheid pulsations. Structures known as partial ionization zones are regions of a star's interior, relatively close to its surface, where the high-temperature ionization (stripping of electrons from atoms) is incomplete. Compressing the gas in these zones releases energy that triggers further ionization and increases opacity.

THE RANGE OF PULSATIONS

This opacity mechanism (now known as the kappa mechanism) offers a good explanation for pulsations in Cepheids and a broad range of other stars. A broad diagonal band on the H–R diagram – the so-called 'instability strip' – is a region where the balance of mass, size and luminosity gives rise

| 1. Slow collapse | 2. Ionization | 3. Expansion | 4. Transparent again |

The kappa mechanism begins with a partial ionization zone transparent to radiation. This reduces radiation pressure so the star's outer layers fall slowly inwards [1]. When temperatures rise far enough, the zone becomes ionized and opaque, trapping radiation [2]. This increases the outward pressure and the star begins to expand [3] until eventually the zone cools and de-ionizes, becoming transparent [4] so that the process can repeat.

to similar zones of partial ionization. Stars on the strip include the so-called 'classical Cepheids', resembling Delta Cephei, and many other forms:

- W Virginis stars. Broadly similar to classical Cepheids but with less mass, these stars have smaller amounts of heavy metals and a distinct period–luminosity relationship.
- RR Lyrae stars. These old stars of 'Population II' are often found in globular clusters (see page 82).
- Delta Scuti stars. Also known as 'dwarf Cepheids' these stars show a similar pattern of variability to Cepheids, but have a much shorter period and are fainter.

Although the kappa mechanism has been successful in explaining many types of variable star, a full understanding of Mira – the finest pulsating star in the sky – remains elusive. Its class of 'long-period variables' are too cool for the kappa mechanism to operate as it does in Cepheids. The mechanism does not seem to alter their overall energy output, but to radically shift it from visible light to the infrared and back again. The most likely explanation at the moment is that their pulsations are created by an *external* opacity mechanism, perhaps a relationship between temperature and the formation of light-absorbing dust in the star's upper atmosphere.

The condensed idea
Many stars vary in brightness, some with a predictable period

29 Supergiants

The brightest stars in the Universe are up to a million times more luminous than the Sun. They range from compact but heavyweight blue supergiants to bloated red supergiants that are less massive, but no less brilliant. Such stellar monsters play a key role in seeding the cosmos with heavy elements.

The search for the brightest and heaviest stars is a perennial pastime for astronomers, but establishing the physics behind them was a crucial breakthrough in our understanding of the Universe as a whole. The term supergiant came from these stars' position on the Hertzsrprung–Russell diagram. A division of different 'luminosity classes' of stars to accompany their spectral types was formalized in the 1940s and 1950s by William Wilson Morgan, Philip C. Keenan and Edith Kellman. Often known as the MK, or Yerkes, classification (see box, page 119), each luminosity class in this system is denoted by a Roman numeral. Normal main-sequence dwarfs are class V, while supergiants are divided into class Ia and Ib. An even brighter class of hypergiants – class O – was added later.

Arthur Eddington suggested as early as the 1920s that there is a luminosity limit above which no star can remain stable against the outward pressure of radiation. Since mass and luminosity are related, this places an upper limit on the mass of stable stars. Until recently, this was thought to be in the low tens of solar masses, but a wide range of factors are now understood to influence the stability of stars and allow them to grow considerably more massive without actually blowing themselves apart in the process

TIMELINE

1843	1867	1943
The luminous blue variable Eta Carinae erupts to briefly become the second brightest star in the sky	Charles Wolf and Georges Rayet identify the first examples of Wolf–Rayet stars	Morgan, Keenan and Kellman introduce the term supergiant for the very brightest stars in their classification system

of formation. Thus, the heaviest star known today is R136a1, a 265-solar-mass behemoth at the heart of a dense young star cluster in the Large Magellanic Cloud galaxy.

VARIED MONSTERS

In the most massive stars, gravity overcomes the natural tendency to expand, keeping them as relatively compact blue supergiants with surface temperatures in the tens of thousands of degrees. Nevertheless, Eddington's essential point that the high luminosities of massive stars render them unstable is borne out in the variety of different supergiants that have so far been studied. Luminous blue variables (LBVs) are highly evolved stars (though just a few million years old thanks to the accelerated lifespan of the most massive stars) and fluctuate violently in size, brightness and surface temperature as they approach the end of their brief lives.

Somewhat less massive than the LBVs are the white supergiants known as Wolf–Rayet stars. First observed in the 1860s, their spectra reveal that they are surrounded by rapidly expanding gas, and they fail to obey the expected relationships of mass, temperature and luminosity. They are a classic example of Eddington's theory in action: stars that begin life with so much luminosity that the high-speed stellar wind blows their surface layers away. Exposing deeper and hotter layers only increases the outward pressure of radiation, creating a runaway effect that can see the star shed a considerable amount of its mass – perhaps tens of Suns' worth – during its brief hydrogen-burning life. This rapid shedding of mass has a significant effect on the way the stars develop in the later stages of their life.

Cooler yellow supergiants are stars that have exhausted their core hydrogen and are expanding towards a red supergiant phase. As they do so, they cross

> THERE SEEMS TO BE A BROAD RELATIONSHIP BETWEEN THE TOTAL MASS OF A CLUSTER, AND THE MOST MASSIVE STAR WITHIN IT.
> Paul Crowther

1954
Hoyle shows how supergiants can generate a variety of fusion processes

1971
Keenan formulates the modern definition of a hypergiant star

2010
Paul Crowther *et al.* identify the heaviest star known, R136a1 in the Large Magellanic Cloud

The interior of an evolved supergiant consists of a huge hydrogen envelope, perhaps with a diameter equivalent to the orbit of Jupiter. At its centre is a relatively small series of shells fusing various nuclei to make elements up to the mass of iron.

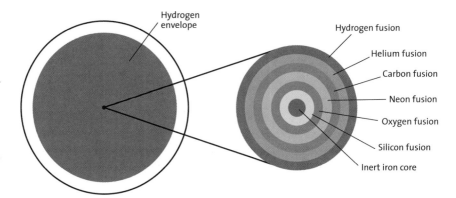

Hydrogen envelope

Hydrogen fusion

Helium fusion

Carbon fusion

Neon fusion

Oxygen fusion

Silicon fusion

Inert iron core

the 'instability strip' of the H–R diagram and become Cepheid variables (see page 112). Thanks to earlier loss of material, they tend to be less massive than the LBVs, with a mass of up to 20 Suns. Red supergiants, meanwhile, are the result of expansion during the final phases of stellar evolution. Analogous to red giants, they form as fusion migrates from the star's core into one or more shell-like layers. These are the largest stars in the Universe by volume, with diameters equivalent to the orbit of Jupiter or even larger. However, only stars of up to 40 solar masses reach this stage; heavier stars come to violent ends in supernovae while they are still LBVs.

ELEMENTAL FURNACES

In 1954, Fred Hoyle outlined the processes going on inside supergiants. He argued that the huge pressures exerted by the outer layers of these stars would compress their cores late in life, making them so hot that fusion processes would not end at the triple-alpha process as they do in Sun-like stars (see page 110). Instead, fusion would continue, forcing together the nuclei of elements, such as carbon and oxygen, together with residual helium to build increasingly heavy elements, such as neon and silicon.

The ongoing process could ultimately create elements up to iron, cobalt and nickel – the heaviest elements whose formation releases more energy than it absorbs. The origins of elements heavier than iron, however, remained a mystery. Then in 1952, husband-and-wife team Geoffrey and Margaret Burbidge discovered a number of unusual stars that seemed to be enriched in these heavier elements. Since they could not have formed through direct fusion, the only alternative way to explain their formation was by slow bombardment of lighter nuclei with individual subatomic neutrons. The Burbidges' work attracted the attention of William Alfred Fowler, and they began to collaborate with both him and Hoyle. The ultimate result was a famous 1957 review paper known (from the acronym of its authors) as B2FH. This outlined for the first time the role not only of nuclear fusion but also of two types of neutron capture, the slow 's-process' and the rapid 'r-process', in the creation of elements within massive stars. Independently, Canadian astrophysicist Alastair G. W. Cameron was following a similar line of research, linking the r-process to supernova explosions (see page 122). This proved to be the crucial final step in explaining the origin of elements within stars.

Luminosity classes

In 1913, German physicist Johannes Stark discovered a phenomenon known as pressure broadening, which makes the absorption or emission lines of spectra associated with a particular gas grow broader when the gas is under higher pressure. This is due to the increased number of collisions with neighbouring gas particles creating slight variations in the overall energy emitted or absorbed by each individual atom. Morgan and Keenan, at Yerkes Observatory, realized that this could be used to estimate the size of a star. Since gases in the photosphere of a small, dense star are at higher pressure than those in the photosphere of a bloated giant, dwarfs should therefore produce broader spectral lines than giants. This was one of the MK system's key innovations, creating a shortcut to identifying stellar luminosities.

Once this independent method of measuring stellar size was combined with information about colour and spectral type, it was possible to directly deduce the luminosities of stars, and by extension work out their likely distance from Earth. This showed for the first time that several apparently unrelated types of stars, with different colours and characteristics, were in fact all highly luminous supergiants.

The condensed idea
Heavyweight stars live fast and die young

30 Supernovae

While stars like the Sun end their lives in the relatively tranquil cosmic smoke-ring of a planetary nebula, stellar heavyweights live fast and die young. They end their lives in the spectacular explosion of a supernova, which may briefly outshine an entire galaxy of normal stars and fling new star-forming material across space.

Supernovae occur at an average rate of about one per century in the Milky Way galaxy, and can sometimes become the brightest objects in Earth's skies. As such, they have been observed and recorded throughout history, most famously in 1054 when the violent death of a star in Taurus lit up the night sky and left behind the shredded cloud of superhot gas known as the Crab Nebula. A later supernova, in 1572, helped shake longstanding assumptions about the unchanging heavens and sped along the Copernican Revolution (see page 5). The rarity of these objects, however, has made them difficult to study – since the invention of the telescope, none are known to have occurred in our galaxy. It was only after the reality of galaxies beyond the Milky Way was accepted in the 1920s that supernova studies really took off, as they began to be spotted and studied in distant galaxies.

TRACKING THE SUPERNOVAE

It took time for astronomers to distinguish supernovae from ordinary novae, which are the occasional flare-ups of otherwise faint stars (see page 129). The first supernova to be recognized as such, in 1934, had actually occurred almost half a century before in the nearby Andromeda Galaxy. It took a breakthrough in distance measurement (see page 146) to make it clear just how violent the

TIMELINE

1921	1934	1941
J.C. Duncan discovers the Crab Nebula is expanding. Knut Lundmark notes its proximity to the nova of 1054	Baade and Zwicky calculate the true brightness of the 1885 S Andromedae nova	Minkowski and Zwicky classify supernovae into distinct types

'S Andromedae' explosion of 1885 had been. Walter Baade and Fritz Zwicky, working at California's Mount Wilson Observatory, calculated that it must have been at least a million times more luminous than the Sun, and coined the term supernova to describe it.

Over the following years, Zwicky, Baade and Rudolph Minkowski carried out an intensive survey of supernovae in other galaxies. Zwicky did the initial search for the appearance of new stars, Baade followed up by measuring the changing brightness of each discovery (building up a model of its 'light curve') and Minkowski concentrated on obtaining spectra. Baade also investigated potential historic supernovae in our own galaxy. He confirmed that the 'new star' of 1572 had in fact been a supernova, and discovered that the Crab Nebula must be a supernova remnant (rather than a planetary nebula), thanks to its rapid expansion rate.

Based on data from more than a dozen individual supernovae, Minkowski and Zwicky produced a classification system in 1941, whose basic features are still used today. Through a combination of spectral line features and differences in the light curve as they fade back to obscurity, they split supernovae broadly into Types I and II, with several subdivisions in each category. However, this division is somewhat misleading, since the objects classed as Type Ia supernovae turned out to have a rather different origin from all the rest (see page 130).

> WE ADVANCE THE VIEW THAT A SUPERNOVA REPRESENTS THE TRANSITION OF AN ORDINARY STAR INTO A NEUTRON STAR CONSISTING MAINLY OF NEUTRONS.
>
> Fritz Zwicky

EXPLODING STARS

Based on the behaviour of S Andromedae, Zwicky and Baade showed as early as 1934 that a supernova explosion involved the conversion of large amounts of mass into pure energy, in accordance with Einstein's famous equation $E = mc^2$. They argued that a supernova represented the transition

1942	1957	1987
Baade measures the expansion rate of the Crab Nebula and links it to the supernova of 1054	The Burbidges, Fowler and Hoyle explain how heavy elements form in supernova explosions	The brightest supernova of recent times, SN 1987A, appears in the Large Magellanic Cloud

between a heavyweight star and something considerably less massive. They also hypothesized the existence of superdense neutron stars (see page 126) as a possible end product of such an explosion. However, it was not until the landmark 'B2FH' paper of 1957 (see page 119) that Margaret and Geoffrey Burbidge, William Fowler and Fred Hoyle outlined the actual processes at work in a typical Type II supernova.

Hoyle's explanation of carbon fusion in stars (see page 110) had convinced him that the interiors of the heaviest stars (more than eight times the mass of the Sun) would build up an onion-like series of fusion shells creating elements all the way up to iron and nickel, beneath an extended hydrogen envelope. Fusing iron, however, would absorb more energy than it released, cutting off the star's power source. Hoyle saw that without outward radiation pressure to support it, once the core's mass inevitably exceeded the Chandrasekhar limit of 1.4 solar masses (see page 126), it would suddenly collapse to form a neutron star (see box, below).

Supernova neutrinos

The formation of a neutron star involves a nuclear reaction in which electrically charged protons and electrons are forced together to form neutrons. In the process, subatomic particles called neutrinos are released as a by-product, and many more are emitted as a way for the neutron star to shed rapidly the excess heat generated by its gravitational collapse. Neutrinos are almost massless and travel very close to the speed of light, so they can emerge from the supernova well before the explosion in its outer layers gathers pace. These fast-moving particles are notoriously hard to detect, but advanced neutrino observatories buried deep underground on Earth offer a useful early warning system for impending supernovae, as well as a means of probing events in and around the core of the exploding star.

COLLAPSE AND REBOUND

Robbed of support, the outer shells fall inwards, only to rebound at the surface of the neutron star, producing a tremendous shock wave. This is the cause of the visible supernova, and the sudden compression and dramatic heating of the star's outer layers unleashes a wave of normally unachievable nuclear reactions. Key among these is the r-process, in which neutrons, produced in abundance by the formation of the neutron star, are captured by heavy nuclei such as iron. Hoyle realized this could produce a wide array of heavy elements in substantial quantities, finally resolving the long-standing problem of their origin.

The B2FH paper convinced many because its predictions matched well with new estimates of cosmic elemental abundance published in 1956 by chemists Hans Suess and Harold Urey (which were based on meticulous measurements of meteorite samples). However, it did not get everything right. Alastair Cameron, working independently, was first to properly explain the importance of the r-process, and it took computer modelling by William Fowler, his student Donald Clayton and Cameron also, to resolve other outstanding problems.

The classic Type II supernova (sometimes called a core collapse supernova) occurs in stars with masses of up to around 40–50 times solar. Type Ib and Type Ic events, which brighten and fade rather differently, involve a similar mechanism, but take place in Wolf–Rayet stars that have shed substantial mass from their outer layers (see page 117). Confusingly, Type Ia supernovae involve an entirely different, and even more spectacular mechanism (see page 129).

Hypernovae

The final stages of really massive stars can be even more dramatic than those of a 'normal' supernova. Monster stars with core masses between 5 and 15 times that of the Sun, collapse to form black holes at their centres (see page 132). These may capture and rapidly gobble up material from the star's outer layers while they are still in the process of exploding. Normally, this stifles the brightness of the original explosion, but if the star is spinning fast enough, the black hole's feeding frenzy will also generate powerful beams of particles moving close to the speed of light. As these interact with the star's exploding outer envelope, they can energize it to ramp up an explosion to ten or twenty times the brightness of a normal supernova. Such 'hypernovae' also release a burst of high-energy gamma rays. Curiously, the most massive cores of all produce neither of these effects – the gravity of the black holes they form is so strong that they swallow up the star before it can fully explode.

The condensed idea
Supergiant stars die
in violent explosions

31 Stellar remnants

At the end of a star's life, it finally sheds its outer layers to expose the burnt-out core that will become its lasting remnant. The exact circumstances in which these final stellar death throes take place, and the kind of object they leave behind in the aftermath, are crucially determined by the star's overall mass.

Astronomers recognize three main types of stellar remnant: in order of increasing density and decreasing size, these are white dwarfs, neutron stars and black holes. The last are the most bizarre objects in the Universe and are covered in more detail in Chapter 33, but the vast majority of remnants are either white dwarfs or neutron stars. Today, we understand that white dwarfs are the end stage of stars with less than eight solar masses, which comprise the overwhelming bulk of our galaxy's stellar population. Neutron stars and black holes are the ghosts of more massive stars that spend their hydrogen-burning lifetimes as supergiants, before dying in spectacular supernovae (see page 120).

WHITE DWARFS

All stellar remnants are far smaller than their progenitor stars, and are therefore much fainter and harder to detect. None are visible to the naked eye, but the first white dwarf to be recorded was noted as a member of the multiple star system 40 Eridani by William Herschel as early as 1783. However, this star's significance was not realized until much later, and as a result the first white dwarfs to be recognized as a significant and unusual class of star were the companions of two of the sky's brightest stars: Sirius

TIMELINE

1862	1926	1931
Clark discovers the small dense white dwarf Sirius B	Fowler describes white dwarfs as collapsed stars supported by degenerate electron pressure	Chandrasekhar calculates the upper limit for a white dwarf's mass

and Procyon. Friedrich Bessel noticed slight shifts in the position of these two nearby stars in 1844 and linked their wobbles to the presence of unseen stars locked in orbit with them. However, Sirius B was not spotted through the telescope until 1862, when it was observed by American astronomer Alvan Graham Clark.

In the early 20th century, astronomers measured the spectra of white dwarfs for the first time and found that they were very similar to 'normal' white stars, but contained enhanced amounts of carbon, nitrogen and oxygen in their atmospheres. Their orbits, meanwhile, indicated that they must carry significant mass despite their faintness. It was clear, then, that these stars were much smaller and denser than those on the main sequence, but nevertheless had extremely hot surfaces. Since their mass could not be held up by radiation pressure as happens in larger stars, something else must prevent white dwarfs from collapsing entirely under their own weight.

> **AS THE CHART FLOWED UNDER THE PEN I COULD SEE THAT THE SIGNAL WAS A SERIES OF PULSES.... THEY WERE 1 1/3 SECONDS APART.**
> Jocelyn Bell Burnell

EXOTIC MATTER

Willem Luyten named these strange little heavyweights 'white dwarfs' in 1922, but an explanation for their unusual properties had to wait until 1926, when the physicist Ralph H. Fowler applied a newly discovered phenomenon of particle physics to the problem. The Pauli exclusion principle states that subatomic electron particles cannot occupy the same state, so that in extreme situations – such as within a collapsed star – they create a 'degenerate electron pressure'. This pressure stops the white dwarf as a whole from crumpling under its own weight, instead creating a superdense star about the size of Earth.

One curious aspect of degenerate electron pressure is that the more matter an object contains, the smaller and denser it will become. Eventually,

1934	1939	1967
Baade and Zwicky predict the existence of neutron stars as supernova remnants	Oppenheimer and Volkoff discover an upper limit to the mass of neutron stars, using earlier work by Tolman	Bell and Hewish discover the first pulsar

Magnetars

An unusual form of neutron star, magnetars may offer a possible explanation for some of the most violent events in the galaxy, so-called 'soft gamma repeaters' that emit powerful periodic bursts of X-rays and even more energetic gamma rays. Magnetars are neutron stars with an unusually slow rotation period, measured in seconds rather than fractions of a second, and an unusually powerful magnetic field generated during the neutron star's initial collapse and sustained by its internal structure. The strength of the field dwindles rapidly over a few thousand years, but while it persists, huge starquakes on the surface of the settling neutron star can lead to sudden rearrangements of the magnetic field, releasing energy that fuels the gamma-ray outbursts.

a threshold is crossed where even electron pressure cannot prevent its collapse. In 1931, Indian astrophysicist Subrahmanyan Chandrasekhar calculated the upper limit of a white dwarf's mass for the first time (about 1.4 solar masses by modern measurements). This important 'Chandrasekhar limit' corresponds to an overall stellar mass of about eight times that of the Sun. Beyond this, Chandrasekhar believed that a white dwarf would inevitably collapse into a black hole.

While Chandrasekhar had got his maths essentially correct, he could not have known that there is an intermediate stage between white dwarf and black hole. The 1933 confirmation of subatomic neutron particles opened up a new area for physicists to explore. It soon became clear that neutrons produce their own degenerate pressure, which operates on even shorter scales than the pressure between electrons. A year later, Walter Baade and his colleague Fritz Zwicky predicted the existence of neutron stars as an end product of supernova explosions (see page 122). They argued that neutron degeneracy could support stars beyond the Chandrasekhar limit, halting their collapse at diameters of just 10 or 20 kilometres (6–12 miles). The tiny size of these objects, it seemed, would make them impossible to observe directly.

COSMIC BEACONS

While neutron stars were undoubtedly interesting hypothetical objects, their presumed invisibility meant that few people bothered to investigate them further. Then, in November 1967, Cambridge PhD researcher Jocelyn Bell came across a curious periodic radio signal from the sky. Lasting just 16 milliseconds and repeating every 1.3 seconds, it came from an object no larger than a planet. At first, it was nicknamed 'LGM-1', a reference to the possibility that it might be a signal from alien 'little green men'. The discovery of similar signals in other parts of the sky soon ruled out this possibility, and the hunt for a cause zeroed in on extreme stellar remnants.

By a remarkable coincidence, Italian astrophysicist Franco Pacini had published a scientific paper just a few weeks before, discussing how the conservation of momentum and magnetic fields would affect a star's collapsed core. Neutron stars, he argued, would rotate extremely rapidly, while their magnetic fields would channel escaping matter and radiation into intense beams emerging from their poles. Pacini and others soon confirmed that Bell had stumbled across just such an object – a cosmic lighthouse now known as a pulsar. However, it was Bell's PhD supervisor Antony Hewish, alongside radio astronomy pioneer Martin Ryle, who was awarded the Nobel Prize in Physics for the discovery.

Quark stars

If a collapsing stellar core has a mass above the so-called Tolman–Oppenheimer–Volkoff (TOV) limit – somewhere between two and three times the mass of the Sun – then even neutron degeneracy cannot create enough pressure to halt its collapse. It used to be assumed that such a core would then immediately collapse into a black hole as its neutrons disintegrated into component particles known as quarks, but modern nuclear physics suggests a possible stay of execution in the form of quark stars. These strange objects are supported by a type of degeneracy pressure between quarks themselves. Quark matter could only remain stable under extreme temperatures and pressures, and might halt collapse at a diameter about half that of a neutron star, around 10 kilometres (6 miles) across. It's also possible that quark matter could create an overdense core inside a neutron star, potentially allowing it to survive beyond the TOV limit.

The condensed idea
Star death leaves behind the strangest objects in the Universe

32 Extreme binary stars

As astronomers investigated realms beyond visible light in the 20th century, a variety of exotic objects were revealed, such as stars emitting strong X-ray and radio signals. The explanation for these strange systems turned out to lie in the interactions between normal stars and stellar remnants.

In a 1941 paper that sought to explain the properties of a curious eclipsing binary (see page 94) called Beta Lyrae, Gerard Kuiper suggested that stars in a binary system might sometimes orbit close enough for material to transfer between them. Kuiper modelled what would happen if one or both stars in such a system overflowed its 'Roche lobe' (the limit where it can hold itself together against the gravitational pull of a neighbour). In the process, he showed that material would not simply be dragged directly from one star to the other, but would build up in an 'accretion disc' above the receiving star's equator. This is particularly likely to happen in systems where a small, dense stellar remnant is accompanied by a bloated red giant with a relatively weak grip on its outer gas layers (a scenario that can arise because stars of different masses age at different rates). In such a situation, the less massive and initially fainter star can end up as the more luminous of the pair, orbited by a small but massive white dwarf, or even (if the star goes supernova) a neutron star or black hole. Combined with the presence of an accretion disc, such a system can produce a range of effects.

CATACLYSMIC VARIABLES
Curiously, binaries involving the least extreme form of stellar remnant (a white dwarf) produce the most violent and spectacular results. As well

TIMELINE

1892	1941	1967
Expanding gas is detected around nova T Aurigae, revealing its explosive nature	Kuiper proposes the existence of contact binaries as an explanation for the star Beta Lyrae	Shklovsky outlines the accretion disc model of X-ray binary stars, used for detecting neutron stars and black holes

as occasional stellar flare-ups known as novae, or 'cataclysmic variables', they can produce the rare and even more impressive outbursts called Type Ia supernovae. The distinction between these two classes of event only became clear in the 1930s, thanks to Fritz Zwicky and Walter Baade's hunt for supernovae in other galaxies (see page 121).

The first nova to be linked to an explosive outburst was T Aurigae, which erupted from obscurity in 1892 to become an easy naked-eye star. Spectroscopic studies suggested that it was surrounded by a shell of rapidly expanding gas, and one early theory suggested that novae were created when stars moved through dense clouds of interstellar gas and heated them to incandescence. The true explanation only emerged from the 1950s onwards, as astronomers established that faded nova systems are typically binaries, with a single visible star orbited by a small, high-mass companion.

Most of the time, the white dwarf in a nova system steadily draws material away from its companion star [1], building up an atmosphere around it via an accretion disc. Occasionally [2] the atmosphere becomes so hot and dense that it explodes in a nuclear firestorm.

By the 1970s, US astronomer Sumner Starrfield and various colleagues were able to establish that the smaller companion stars in nova systems were white dwarfs, and to develop a 'thermonuclear runaway' model to explain what was happening. According to this theory, novae only occur in tight binary systems where the system's larger star overflows its Roche lobe, allowing the white dwarf to tug material away from its extended gaseous envelope. Captured gas from the accretion disc piles up to create a hydrogen layer around the dwarf itself, compressed by the powerful gravity and also heated by the incandescent surface below. Eventually, conditions in the hydrogen atmosphere become so extreme that nuclear fusion takes hold, burning its way through the atmosphere in a runaway reaction that may last for several weeks. Once the

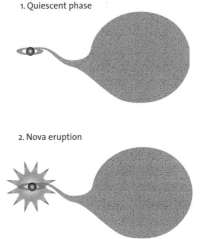

1. Quiescent phase

2. Nova eruption

1971–4	1973	1974
Starrfield and colleagues explain novae as thermonuclear explosions associated with white dwarfs in contact binary systems	John Whelan and Icko Iben, Jr. explain Type Ia supernovae through the sudden collapse of a white dwarf stars	Warner explains the origin of dwarf nova outbursts

Dwarf novae

Some nova outbursts occur on a much smaller scale than usual, and in semi-regular periods that range between days and years. These 'dwarf novae' (often called U Geminorum stars, after the prototype discovered in 1855) involve the same kind of mass-transfer binary system seen in the much brighter 'classical novae'. In 1974, British astronomer Brian Warner first outlined how they generate their outbursts through a quite different mechanism. Captured material building up in the accretion disc reaches such high densities that it becomes unstable, triggering a sudden collapse onto the surface of the white dwarf and a dramatic explosion that slowly fades before the system returns to normal. Several hundred U Geminorum stars have been discovered, revealing a clear pattern between the intensity and frequency of their eruptions: the longer the wait between outbursts, the brighter they eventually are. Some astronomers are therefore keen to investigate the possibility of using dwarf novae as standard candles – a means of calibrating distances across our galaxy and others.

supply of hydrogen is exhausted, the nova fades away, but the process may re-establish itself and eventually repeat in a so-called 'recurrent nova'. Between outbursts, radiation from material entering the accretion disc causes the system's overall light output to flicker in a very distinctive way.

DISINTEGRATING STARS

The intensity of nova outbursts and the period between repeated eruptions depends on the exact dynamics of the system, so no two novae are identical. Yet the same cannot be said for the cataclysmic variable's bigger brother, the Type Ia supernova, and in fact a key breakthrough in our modern understanding of the Universe hinges on the fact that the intensity of these spectacular explosions is always the same.

Type Ia supernovae develop from nova systems where a white dwarf is close to the Chandrasekhar limit of 1.4 solar masses (see page 126). Astronomers used to assume that if enough mass accumulated in the dwarf's atmosphere, it simply underwent a sudden and violent collapse into a neutron star, but recent research has suggested that before this can happen, the growing internal pressure triggers a fresh wave of fusion in the carbon and oxygen locked inside. Because its material is degenerate, the white dwarf cannot expand in the same way as a normal star, so its core temperature soars to billions of degrees and fusion runs out of control. The conditions for degeneracy are finally broken in a sudden and dramatic explosion that completely destroys the star, and whose peak brightness is about 5 billion times more luminous than the Sun. Because Type Ia supernovae always involve the same amount of mass being converted into energy, cosmologists can use them as standard candles for measuring the distance to remote galaxies (see page 184).

X-RAY BINARIES

If the invisible star in a binary system is a neutron star or black hole, the results can be very different. Rather than producing intermittent visible outbursts, matter falling into the accretion disc is torn to shreds and heated by extreme tidal forces thanks to the stellar remnant's much more intense gravitational field. At million-degree temperatures, parts of the disc become strong but variable sources of high-energy X-rays – a mechanism used by Iosif Shklovsky in 1967 to explain why some visible stars are apparently also bright X-ray sources.

The vast majority of neutron stars so far identified are known from either X-ray binaries or the pulsar mechanism (see page 127), and until very recently, X-ray binaries have been the only means of locating stellar-mass black holes (see page 127). However, there is also no theoretical reason why binaries should not exist containing two neutron stars or even two black holes – and such systems are indeed thought to make up a substantial proporition of all extreme binaries. The powerful gravity between stellar remants in either type of binary creates strong tides that send them spiraling toward each other on an inevitable collision course, whose final moments generate bursts of gravitational waves (see page 191). While the union of two black holes should produce no outward explosion, mergers between neutron stars may be responsible for enormous gamma-ray outbursts, and some have argued that they could also offer another means of forming the heaviest elements in the Universe.

> **SOMETHING THE SIZE OF AN ASTEROID IS A BRILLIANT, BLINKING SOURCE OF X-RAYS, VISIBLE OVER INTERSTELLAR DISTANCES. WHAT COULD IT POSSIBLY BE?**
> Carl Sagan

The condensed idea
Stars in binary systems can have radically different life cycles

33 Black holes

The idea of objects with so much mass that light cannot escape their gravity has been around for a surprisingly long time, but understanding the physics involved in such strange objects is no easy task, and tracking down an object that gives out no light can be even tougher.

In 1915, Albert Einstein published his theory of general relativity. This model unified space and time in a four-dimensional spacetime 'continuum' that can be distorted by large accumulations of mass, giving rise to the effect we experience as gravity (see page 190). He described the theory through 'field equations', and just a few months later, Karl Schwarzschild used these to investigate how spacetime would become distorted around a large mass occupying a single point in space.

Schwarzschild showed that if any mass was compressed to beneath a certain size (now known as its Schwarzschild radius), the description offered by Einstein's equations would break down: in mathematical language, the object would become a singularity. Furthermore, the speed required to escape from the object's gravity (what we now call the escape velocity) would exceed the speed of light. Since this is the fastest speed in the Universe according to relativity, such an object would effectively be inescapable.

Arthur Eddington, who had already done much to champion Einstein's theory (see page 191), considered such compressed objects in his 1926 book on stellar structure, and made a significant refinement to the basic idea.

TIMELINE

1783	1915	1926	1931
Michell predicts the existence of dark stars with such high gravity that light cannot escape	Schwarzschild predicts the existence of black holes from his analysis of general relativity	Eddington shows how singularities would red shift the light around them	Chandrasekhar argues that singularities can result from the collapse of the most massive stellar cores

Since the speed of light is constant, light from such a superdense star cannot actually slow down. Instead, Eddington argued, it must lose energy by being increasingly red shifted to longer wavelengths. When the star is compressed below its Schwarzschild radius, its light is effectively red shifted into invisibility.

FROM THEORY TO REALITY

However, Schwarzschild's strange objects remained purely theoretical until 1931, when Subrahmanyan Chandrasekhar suggested that they would inevitably result from the collapse of a stellar core containing more than 1.4 solar masses of material (see page 126). Chandrasekhar argued that there was no means for such a star to generate sufficient pressure to balance against its own gravity. He did not anticipate the later discovery of neutron stars, but in 1939 Robert Oppenheimer and his colleagues showed that even these superdense stars have an upper mass limit of around three solar masses. Oppenheimer argued that as the collapsing star passed the Schwarzschild radius, the passage of time came to a halt, and so for a while these counterintuitive objects became known as frozen stars.

A new era in the study of black holes began in 1958, when US physicist David Finkelstein redefined the Schwarzschild radius as an 'event horizon'. Within

Predicting dark stars

In 1783, English clergyman and astronomer John Michell presented a remarkably prescient paper to the Royal Society in London. At the time, most scientists followed Isaac Newton's corpuscular theory of light, in which light consisted of tiny particles moving at high speeds. Michell reasoned that such particles would be affected by gravity, and therefore it was theoretically possible for a star to have such powerful gravity that the speed required to escape from it would overcome the speed of light. In such a case, he argued, the result would be a 'dark star' – an object that emitted no radiation but could nevertheless be detected by its gravitational influence on visible objects, for instance, if such a star existed in a binary system. Michell's paper was an impressive prediction of the black hole phenomenon, but his work was neglected until the 1970s, by which time black holes had been discovered by other means.

1958	1963	1969	1973	2015
Finkelstein develops the idea of the event horizon	Roy Kerr models rotating black holes, the type most likely found in nature	Lynden-Bell proposes super-massive black holes as a possible explanation for quasars activity	Webster, Murdin and Bolton demonstrate that Cygnus X-1 is a likely black hole	A black hole merger is detected for the first time through gravitational waves

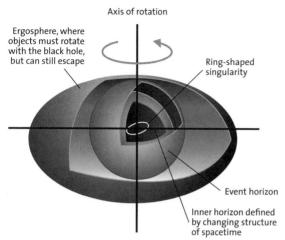

Axis of rotation

Ergosphere, where objects must rotate with the black hole, but can still escape

Ring-shaped singularity

Event horizon

Inner horizon defined by changing structure of spacetime

Rotating black holes are likely to be the most common form in nature. According to New Zealand mathematician Roy Kerr's 1963 analysis, they have several features not present in static black holes.

this boundary the star's collapse continued to form an infinitely dense point in space (the true singularity), but from an outsider's point of view no information could escape the horizon – and anything that crossed that boundary was doomed to a one-way trip.

BEYOND THE EVENT HORIZON

During the 1960s and early 1970s, cosmologists looked more deeply into the information properties of these strange objects, discovering that their properties were influenced only by the mass of material they contained, its angular momentum and its electric charge. According to the 'no hair theorem', all other information is irretrievably lost. The term 'black hole' to describe these objects was coined by journalist Ann Ewing in a 1964 report and gained popularity when it was adopted by physicist John Wheeler a few years later.

In 1969, British astrophysicist Donald Lynden-Bell suggested for the first time that black holes might not be limited to stellar-mass objects, arguing that the disappearance of matter into an enormous 'Schwarzschild throat' with the mass of millions of Suns might drive the strange and violent activity seen in the hearts of active galaxies (see Chapter 38). Such an object, now called a supermassive black hole, could be initiated by something as simple as the collapse of an enormous cloud of interstellar material in a young galaxy. In 1971, Lynden-Bell and his colleague Martin Rees went as far as suggesting that a dormant black hole formed the gravitational anchor at the heart of our own galaxy (see page 140).

DETECTING BLACK HOLES

In 1974, a young physicist called Stephen Hawking made his reputation by showing that despite no radiation escaping from within, quantum physics effects would cause a black hole to *generate* low-intensity radiation at its event horizon, with a wavelength related to its mass. Nevertheless, this 'Hawking radiation' is so weak as to be undetectable, so black holes are, to all intents and purposes, invisible.

Fortunately for astronomers, however, conditions around black holes are so extreme that they produce other effects that can be detected. Specifically, X-rays are emitted from material falling into a black hole, as tidal forces from the enormous gravity tear it apart and heat it to million-degree temperatures. In the 1960s, several astronomical X-ray sources were discovered by rocket-borne instruments, with hundreds more found after the launch of the first dedicated X-ray astronomy satellite, Uhuru, in 1970. Many of these proved to be clouds of superhot gas within distant galaxy clusters (see page 157), but some were compact and appeared to be associated with visible stars in the Milky Way.

> **THE BLACK HOLES OF NATURE ARE THE MOST PERFECT MACROSCOPIC OBJECTS THERE ARE IN THE UNIVERSE.**
> S. Chandrasekhar

The most likely scenario to explain these bright, rapidly variable sources was a so-called 'X-ray binary'. These are compact stellar remnants, that pull material away from a visible companion star (see page 131). Usually, such systems involve neutron stars, but in 1973, however, British astronomers Louise Webster and Paul Murdin, along with Canadian Thomas Bolton, investigated the bright X-ray source Cygnus X-1 and measured the Doppler shift of light from its visible counterpart, a blue supergiant star. This revealed that the star is locked in orbit around an unseen companion with more than eight times the mass of the Sun. Such an object could only be a black hole. This basic idea of detecting a black hole through its influence on a companion star has since been used to detect several similar systems.

The condensed idea
There is no escape from the densest objects in the Universe

34 The Milky Way galaxy

The Milky Way is a band of pale light that wraps itself around the night sky. Celebrated since prehistoric times, its deeper nature was only revealed after the invention of the telescope, and its identity as a vast spiral system of stars was only defined in the 20th century.

Unsurprisingly, the Milky Way was one of the first targets for Italian astronomer Galileo Galilei, who turned his primitive telescope towards it in January 1610. Discovering that it was studded with countless stars that could not be seen with the naked eye, he concluded that the entire band comprised countless further stars beyond the reach of his instrument. Going even further, he argued that fuzzy cloudlike 'nebulae' were also made up of distant stars (a conclusion that is correct in some cases but not all).

It was only in 1750, however, that English astronomer Thomas Wright argued that the Milky Way must be a vast rotating cloud of stars, confined by gravity to a single plane with a structure broadly similar to our own solar system. Five years later, Immanuel Kant also discussed a disc-like galaxy, suggesting, with considerable prescience, that it was just one of many 'island universes', some of which were visible over immense distances as nebulae.

MAPPING THE MILKY WAY
William Herschel made the first attempt to chart the Milky Way in the 1780s. He counted the number of stars in different areas of the sky, and assumed that all stars had the same inherent brightness, so that their apparent

TIMELINE

c.1000–1300	1610	1750	1785
Various Islamic astronomers argue that the Milky Way is made up of the light from countless stars	Galileo makes the first telescopic study of the Milky Way and discovers many new stars	Wright makes the first estimate of the shape of the galaxy based on the distribution of stars	Herschel publishes the first map of the Milky Way

magnitude was a direct indication of their distance. This led him to map our galaxy as an amorphous blob with the Sun close to the centre. More than a century later, Dutch astronomer Jacobus Kapteyn led a far more exhaustive effort to repeat Herschel's work, using more powerful instruments and a full range of astronomical data to estimate the true brightness of the stars. However, Kapteyn's survey, finally published in 1922, reached more or less the same conclusions, proposing a lens-shaped galaxy some 40,000 light years in diameter with the Sun close to its centre.

Ironically, by the time Kapteyn published his work, a discovery had already been made that would undermine his view of the galaxy. In 1921, Harlow Shapley compiled his own survey of the dense globular star clusters found in some parts of the sky (see page 82). He concluded that they were loosely clustered around a distant region of space in the direction of the constellation Sagittarius. This, Shapley believed, was the true centre of the Milky Way, with our solar system lying far away in the outer reaches of its broad disc.

> **THE MILKY WAY IS NOTHING ELSE BUT A MASS OF INNUMERABLE STARS PLANTED TOGETHER IN CLUSTERS.**
> Galileo Galilei

One thing Shapley got wrong, however, was his estimate of the true diameter of the Milky Way. Based on erroneous estimates of the distance to globular clusters, he assumed it was an enormous 300,000 light years across. This began to be corrected from 1927 when Jan Oort set out to demonstrate a theory (proposed shortly before by Swede Bertil Lindblad) that stars rotated at different speeds depending on their distance from the centre of the galaxy. Oort's careful measurements allowed him to develop a formula for calculating this 'differential rotation', and proved that the solar system was about 19,000 light years from the centre of a galaxy some 80,000 light years across. This is just a slight underestimate from the modern values of 26,000 and 100,000 light years, respectively.

1921	1927–40	1930	1956	2005
Shapley identifies the rough centre of the Milky Way from the distribution of globular clusters	Oort determines the scale of the galaxy from stellar motions	Robert Trumpler catalogues open clusters in the Milky Way and identifies light-absorbing dust between the stars	Oort confirms the Milky Way's spiral structure from mapping of hydrogen clouds	Infrared observations confirm that our galaxy is a barred spiral

Creating spiral arms

Our galaxy's differential rotation means that its spiral arms cannot possibly be permanent physical structures – if they were, the faster rotation of regions near the galactic hub would cause them to 'wind up' and disappear within just a few rotations. Instead, the spiral structure must be constantly regenerated

Today, we understand that spiral arms are regions of conspicuous star formation within a disc of stars, gas and dust that surrounds the hub. Individual objects move in and out of these regions over tens of millions of years – stars slow down and bunch together like cars moving through a traffic jam, while interstellar clouds are compressed to trigger the creation of new stars. The brightest and most massive of these have such short lifespans that they age and die before getting a chance to leave the nursery regions and join the general disc population.

But how does the spiral 'traffic jam' region itself arise? The best available explanation, proposed by Chia-Chiao Lin and Frank Shu in the late 1960s, is known as the density wave theory. It relies on the fact that all objects orbiting the galactic centre follow elliptical, rather than perfectly circular, orbits, and move more slowly near the outer edges of these orbits. When an outside influence, such as interaction with a small satellite galaxy, tugs these orbits into alignment, the result is a spiral zone where stars and other material are more likely to be found.

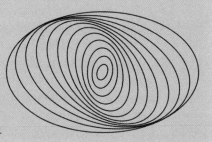

This schematic shows how spiral areas of higher density arise naturally when a number of elliptical orbits are influenced by slightly different amounts, as during a galaxy close encounter.

SPIRAL ARMS

The first confirmation of the Milky Way's spiral structure developed from William W. Morgan's attempts to map the distribution of open clusters in the early 1950s. Morgan identified three distinct chains of clusters that he suggested might be fragments of spiral arms, and his discovery was confirmed a few years later when Jan Oort used radio observations to chart the distribution of clouds of neutral atomic hydrogen across the galaxy. Radio signals emitted by hydrogen at a wavelength of 21 cm (8.3 in) penetrated through intervening star clouds and dust lanes, allowing Oort to map the galaxy on a far greater scale than Morgan had.

As astronomers got to grips with these new techniques, a picture emerged of a spiral with four major arms and several minor structures called spurs running between them (one of these, the Orion–Cygnus Spur, is the closest to our own solar system). The Milky Way was generally agreed to be a 'normal' spiral with an ovoid central hub, but in the 1970s new radio maps began to suggest a bar-shaped zone of star formation extending to either side, and a huge ring of starbirth surrounding the galactic centre at a radius of about 16,000 light years. From intergalactic space, this ring might well be our galaxy's dominant feature.

In 2005, infrared observations from NASA's Spitzer Space Telescope confirmed the existence of the central bar, tracing the distribution of red giants across a 28,000-light-year span and confirming beyond doubt that the Milky Way is actually a *barred* spiral (see page 146). Galaxies of this type have two major spiral arms (one emerging from either end of the bar), and in 2008 astronomer Robert Benjamin of the University of Wisconsin used Spitzer observations to trace a concentration of cool red giants into two arms. However, in 2013, a new radio survey re-established the separation of new star-forming regions and young stars into *four* major arms.

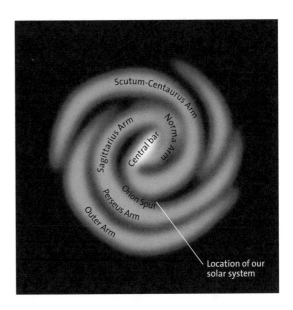

A simplified map of the Milky Way shows the position of our solar system and the galaxy's major features.

Further studies will undoubtedly be needed to resolve this discrepancy between old and young stars, but the solution may prove to be connected with our galaxy's ongoing series of collisions with the smaller Sagittarius Dwarf Elliptical galaxy. According to computer models published in 2011, this small galaxy – currently some 50,000 light years away on the far side of the galactic centre – is almost certainly responsible for shaping the Milky Way's current spiral structure.

The condensed idea
Our galaxy is a spiral of stars, with the Sun far from its centre

35 Heart of the Milky Way

The central regions of our galaxy lie 26,000 light years away in the constellation of Sagittarius. Dense intervening star clouds block the core itself completely from visual observation, but breakthroughs in radio- and space-based astronomy have revealed the presence of a slumbering monster at the heart of the galaxy: a black hole with the mass of four million Suns.

Following Harlow Shapley's discovery of the true centre of our galaxy in 1921 (see page 137), astronomers naturally turned their attention to studying this intriguing area of the sky. Despite the limited techniques available to them at the time, they were soon able to compare the Milky Way's structure to that of the so-called spiral nebulae, which in the 1920s had newly been confirmed as galaxies in their own right. They soon concluded that the centre of our galaxy is marked by a 20,000-light-year-wide bulge of red and yellow Population II stars (see box). But what could have brought this enormous star cloud together in the first place?

The idea that galactic cores might conceal 'supermassive' black holes with the mass of millions of Suns emerged from attempts to explain quasars and other active galaxies in the 1960s and 1970s (see page 152). In 1971, Donald Lynden-Bell and Martin Rees suggested that dormant supermassive black holes might lie at the centre of all galaxies, including the Milky

TIMELINE

1921	1933	1971	1974
Shapley locates the centre of our galaxy in a distant part of Sagittarius	Jansky identifies radio emissions from the galaxy's central regions	Lynden-Bell and Rees suggest there is a supermassive black hole at the heart of the Milky Way	Brown and Balick identify the compact radio source Sagittarius A*

Way, acting as the gravitational hub around which the entire system spins. With no way of seeing through the intervening star clouds in visible light, and with space-based observations still in their infancy, the initial evidence corroborating Lynden-Bell and Rees' ideas emerged from the field of radio astronomy.

SIGNALS FROM THE CORE

The first radio telescope was a makeshift array of antennae very different from the later parabolic dishes. Built by physicist Karl Jansky at the Bell Telephone Laboratories in New Jersey around 1930, it had only crude directional capabilities, but these were enough for Jansky to identify a radio signal from the sky that appeared to rise and set daily. At first, the signal seemed to be matching the motion of the Sun, but over several months,

Stellar populations

The idea of two distinct stellar populations was first floated by Walter Baade based on his studies of the nearby Andromeda Galaxy, and subsequently applied to stars elsewhere, including those in our own galaxy. Stars of Population I are found in the discs and arms of spiral galaxies. They are relatively young, have a range of colours and a fairly high 'metallicity' (proportion of elements heavier than hydrogen and helium), which allows them to shine through the CNO cycle (see page 75). Population II stars, in contrast, are found mostly in the central bulges of spiral galaxies, and in globular clusters and elliptical galaxies (see pages 82 and 145). They are individually faint, generally less massive than the Sun and overwhelmingly red and yellow. A lack of metals limits their hydrogen fusion to the proton–proton chain (see page 74) and ensures they have long, unspectacular lives. Population II stars are generally thought to be the oldest in the Universe today, with some still surviving from the first billion years after the Big Bang.

Jansky note that it drifted away: rising and setting a little earlier each day, its motion was actually matching the rotation of the stars. By 1933, he was able to announce the detection of radio waves coming from the Milky Way, and strongest in the direction of Sagittarius.

This central radio source, later designated Sagittarius A, remained an amorphous, diffuse blob until the 1960s, when astronomers were finally

1998	2008-9	2009	2015
Ghez *et al* confirm the presence of a black hole from the rapid motion of stars around Sagittarius A*	Astronomers constrain the black hole's mass to around 4.2 million solar masses	Stefan Gillessen *et al.* discover large amounts of unseen material close to the central black hole	X-ray telescopes detect the destruction of an asteroid entering the black hole

able to resolve it in finer detail. It turned out to be divided into distinct eastern and western units: the eastern half is today recognized as a supernova remnant, while 'Sagittarius A West' is a curious three-armed spiral structure. Then in 1974, Robert Brown and Bruce Balick pinned down a third distinct element. This was a much more compact source within Sagittarius A West, which was subsequently named Sagittarius A*. Astronomers immediately speculated that this object might mark the huge concentration of mass at the exact centre of the galaxy as a whole, and this was confirmed in 1982 by precise measurements of its motion – or more accurately, its distinct *lack* of motion.

THE KEY TO PROVING THAT THERE'S A BLACK HOLE IS SHOWING THAT THERE'S A TREMENDOUS AMOUNT OF MASS IN A VERY SMALL VOLUME.

Andrea M. Ghez

The 1970s and 1980s also saw the arrival of other methods for looking beyond the intervening star clouds. Infrared satellites proved particularly useful in identifying massive open star clusters around the central region. One of these, known as the Quintuplet, proved to host a truly enormous stellar monster known as the Pistol Star. Shining an estimated 1.6 million times brighter than the Sun, this is just the largest among many stellar giants in both the Quintuplet and the more massive nearby Arches cluster (only discovered in the 1990s).

Although both of these clusters are located tens of light years from Sagittarius A*, the presence of such short-lived monster stars undermined assumptions that the galactic hub would be home only to subdued and long-lived Population II dwarf stars. Instead, it is clear that the core regions have been sites of active star formation over the past few million years.

ORBITING A MONSTER

Although not a match for the Arches and Quintuplet, another significant cluster of high-mass stars surrounds Sagittarius A* itself. Discovered in the 1990s and unassumingly known as the 'S-Star cluster', these stars have played a key role in proving the existence of the galactic black hole and constraining its properties.

Doppler shifts reveal that the cluster's stars are all moving at speeds of hundreds of kilometres per second or even more. Following elliptical orbits around an unseen central body, it's possible to track their shifting position over a matter of years, constraining the size of the massive object that anchors the cluster and the entire Milky Way. One star in particular, a 15-solar mass giant designated S2, has been tracked continuously since 1995. It follows a roughly 15.6-year orbit around Sagittarius A*, with a closest approach roughly four times the distance between the Sun and Neptune. Analysis of S2's orbit and that of S102 – an even closer star discovered in 2012 – confirms the existence of an unseen object with roughly 4 million times the mass of the Sun in a region substantially smaller than Earth's orbit. This object can only be a black hole.

Light sleeper?

In the past decade, studies have found that our galaxy's central black hole has been active in the relatively recent past. Orbiting X-ray telescopes occasionally observe powerful flares from the galactic centre, most likely caused when small objects such as asteroids stray too close and are shredded and heated by the black hole's immense gravity. They have also found 'light echoes' – glowing clouds of emission created as X-rays from a more violent event a few decades ago illuminate gas clouds about 50 light years from the black hole.

Because anything that strays too close will be pulled rapidly to its doom, most astronomers assumed that the central black hole would have cleared out its immediate surroundings and subsided into inactivity. The only things left would be a few risk-taking stars like S2 and S102 orbiting nearby, and a slow but steady drift of gas onto the black hole generating the radio signals of Sagittarius A*. It was something of a surprise, then, when a study published in 2009 suggested that the region where S2 orbits is packed with another million Suns' worth of material, believed to be distributed among otherwise-undetectable faint stars and stellar remnants. In such a crowded environment, the central black hole may not be as dormant as previously thought.

The condensed idea
A supermassive black hole lies at the centre of the Milky Way

36 Types of galaxies

Edwin Hubble's 1924 discovery that many of the sky's nebulae are independent galaxies far beyond our own opened up an entirely new field of astronomy. Comparisons could now be made between the Milky Way and these other systems, and it was immediately clear that some types of galaxy are very different.

Today's system of galaxy classification has been modified and amended many times, but astronomers still recognize the five major galaxy types identified by Hubble in his 1936 book, *The Realm of the Nebulae*. These are spirals, barred spirals, ellipticals, lenticulars and irregulars.

Spirals have a bulging nucleus of older Population II stars (see page 141), out of which emerge spiral arms highlighted by star-forming regions and bright clusters of luminous, short-lived stars of Population I. In 1939, Horace Babcock used spectroscopic measurements of the Andromeda Galaxy to confirm that the stars within spirals rotate at different rates depending on their distance from the centre, an idea first proposed by Bertil Lindblad in 1925 (see page 137). Typical rotation periods midway from the centre to the edge are around 200 million years. As well as normal spirals, Hubble identified a large group of galaxies in which a straight bar crosses the nucleus, with the spiral arms emerging from the ends. Barred spirals actually account for about two-thirds of spirals in the nearby Universe, including our own Milky Way.

Together, spirals and barred spirals account for about 60 per cent of bright galaxies in the present epoch, though this number has undoubtedly changed

TIMELINE

1924	1936	1937	1939
Hubble confirms that spiral nebulae are galaxies far beyond the Milky Way	Hubble outlines a broad classification of galaxy types	Shapley discovers the first of the abundant dwarf spheroidal galaxies	Babcock confirms the differential rotation of stars across spiral galaxies

over time. They range in size from a few tens of thousands to around half a million light years across, though systems bigger than the Milky Way's 100,000-light-year diameter are very rare. Hubble subdivided both types of spiral according to how tightly wound their arms appeared. There are several other important distinctions, among which perhaps the most significant separates 'grand design' spirals, with sharply defined spiral arms, from 'flocculent' spirals, in which star formation is a more diffuse and patchy affair. The difference between the two is thought to be determined by the relative influence on star formation of large-scale factors such as a spiral density wave (see page 138) and local factors such as supernova shock waves.

> **THE HISTORY OF ASTRONOMY IS A HISTORY OF RECEDING HORIZONS.**
> Edwin Hubble

ELLIPTICALS AND LENTICULARS

Hubble's third major class of galaxies are ellipticals. These are ball-shaped clouds of red and yellow stars in orbits that are not only elongated, but inclined at a wide range of angles. Unlike spirals, ellipticals are largely lacking in the interstellar gas clouds required to form new stars. Any short-lived, massive blue and white stars have long since aged and died, leaving behind only the more sedate, low-mass Population II stars. The lack of gas is also responsible for their chaotic structure – collisions between gas clouds have a natural tendency to create discs on any scale from solar systems to galaxies, and the gravitational influence of these discs in turn flattens stars' orbits. Close encounters between stars – another mechanism for 'averaging-out' orbits – are rare and so ellipticals take a range of forms between perfect spheres and elongated cigar shapes. They vary much more widely in size than spirals, with diameters ranging from a few thousand to a few hundred thousand light years across. They account for about 15 per cent of all galaxies today, but the largest examples are only ever found in dense galaxy clusters – a trait that offers an important clue to their origins (see pages 150 and 157). Hubble arranged spirals, barred spirals and ellipticals on his famous 'tuning

1944	**1959**	**1964**
Baade identifies two stellar populations in the Milky Way and other galaxies	Gérard de Vaucouleurs introduces a widely used extension to the Hubble system	Lin and Shu propose their density wave theory to explain spiral arms

Discovering other galaxies

The nature of spiral nebulae – established as distant clouds of stars through spectroscopic studies – was a subject of fierce astronomical debate in the early 20th century. Were they relatively small systems in orbit around a Milky Way that effectively encompassed the entire Universe, or were they large and distant galaxies in their own right, implying a much greater scale to the Universe?

The so-called 'Great Debate' was finally settled in 1925 through the assiduous work of Edwin Hubble, building on that of Henrietta Swan Leavitt (see page 113) and Ejnar Hertzsrprung. Hubble combined Leavitt's discovery of a relationship between period and luminosity in Cepheid variable stars with Hertzsprung's independent determination of the distance to several nearby Cepheids. This allowed him to use the period of Cepheids to estimate their intrinsic luminosity, and therefore (by comparison with their apparent brightness in Earth's skies) their likely distance from Earth. Over several years Hubble used the 2.5-metre (8.2-ft) telescope at California's Mount Wilson Observatory to locate Cepheids in some of the brighter spiral nebulae and monitor their brightness. By 1924, he was able to confirm that the spiral nebulae were independent systems millions of light years beyond the Milky Way – the first step in an even greater discovery (see page 160).

fork' diagram (see opposite), with an intermediate type of galaxy called the lenticular at the junction between the fork's two arms. Lenticulars resemble 'armless spirals'. They have a central bulge of stars surrounded by a disc of gas and dust, but little sign of ongoing star formation to create spiral arms. As Hubble correctly guessed, they are thought to mark a key stage in the evolution of galaxies from one form to another.

IRREGULAR GALAXIES AND DWARF SPHEROIDALS

Smaller than spirals, Hubble's irregular galaxies are amorphous clouds of stars, gas and dust, often rich in bright young stars and distinctly blue in colour. They are thought to account for about a quarter of all galaxies, though they are generally smaller and fainter than either spirals or ellipticals, making them hard to observe. Fortunately, two of our closest galactic neighbours – the Large and Small Magellanic Clouds – are irregular, so the type is well studied.

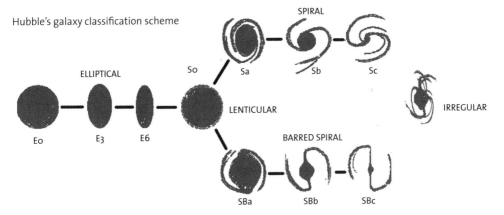

Hubble's galaxy classification scheme

SPIRAL

ELLIPTICAL

So

Sa

Sb

Sc

LENTICULAR

IRREGULAR

Eo

E3

E6

BARRED SPIRAL

SBa

SBb

SBc

Hubble divided irregulars into two classes: 'Irr I' galaxies that show some internal structure, and 'Irr II' galaxies that are entirely shapeless. The structure within larger Irr I galaxies can include traces of central bars or poorly defined spiral arms. Images of the distant, early Universe from the Hubble Space Telescope show irregular galaxies were much more abundant in the past, and support the idea that spirals originated from mergers.

One final significant group of galaxies, even smaller and fainter than the irregulars, are the dwarf spheroidals. These small spherical or elliptical clouds of stars were discovered by Harlow Shapley in 1937, and have no obvious nucleus and a very low surface brightness. Despite an outward similarity to elliptical galaxies, dwarf spheroidals seem to contain a more complex mix of stars, as well as large amounts of unseen 'dark matter' whose gravity holds their sparse visible elements together. They make up more than two thirds of all galaxies in the vicinity of our own, but are impossible to detect over larger distances.

The condensed idea
Galaxies come in several very different forms

37 Colliding and evolving galaxies

While the distances separating galaxies are vast compared to our everyday scales, they are relatively small compared to the size of galaxies themselves. This makes collisions and close encounters between galaxies a surprisingly common event, and one that plays a key role in the evolution of galaxies.

Once the true nature of galaxies became clear in the 1920s, astronomers soon discovered that many galaxies lying near each other in the sky are genuinely close to each other in space. Swedish astronomer Erik Holmberg carried out pioneering work in this area as early as 1937, and in 1941 was the first person to consider what would happen if two galaxies collided. To do this, he used a primitive analogue computer, built from dozens of light bulbs whose varying intensities could show concentrations of stars. Holmberg's work revealed several important effects: he showed how the approaching galaxies would bring about tidal forces within each other, triggering waves of star formation, while slowing down their overall motion through space so that they would ultimately coalesce and merge together.

Despite this, galaxy collisions were disregarded as rare accidents until 1966, when Halton Arp published his *Atlas of Peculiar Galaxies* – a catalogue outlining the wide variety of galaxies that did not fit into Edwin Hubble's neat classification scheme (see page 147).

TIMELINE

1941	1951	1966
Holmberg models the events associated with hypothetical galaxy collisions	Lyman Spitzer, Jr. and Walter Baade suggest that collisions might be a mechanism for transforming galaxies from one type to another	Halton Arp publishes his *Atlas of Peculiar Galaxies*

MODELLING MERGERS

At around the same time, Estonian brothers Alar and Jüri Toomre applied new supercomputer technology to the problem of mergers. They produced similar results to those of Holmberg, but with much more detail. In some cases, they even simulated the collisions of specific galaxies. The Antennae galaxies, for example, are a pair of colliding spirals about 45 million light years away in the constellation of Corvus: as they approached each other, tidal forces 'unwound' their spiral arms, creating two long streamers of stars that extend across intergalactic space. Hubble Space Telescope images since the 1990s have revealed intense star formation in the main bodies of these galaxies, while X-ray images show that the entire system is now surrounded by a halo of hot gas.

Despite the spectacle, it seems that collisions between individual stars are rare. The more diffuse clouds of gas and dust collide head-on, creating abundant new stars in an event known as a starburst. Shock waves tearing through the colliding material heat it substantially. Meanwhile, supernova explosions from massive but short-lived stars created in the starburst heat the gas even more, raising its

Super star clusters

One of the most spectacular results of galaxy interaction is the formation of star clusters on a scale that dwarfs normal open or globular systems (see page 82). So-called super star clusters are the component elements of a broader starburst, created as powerful tidal forces trigger the gravitational collapse of huge interstellar gas clouds. The most prominent such cluster in Earth's skies is R136, a particularly dense open cluster in the Large Magellanic Cloud that is home to the heaviest stars so far discovered (see page 117). However, at least two super star clusters are now recognized in the Milky Way itself.

Super star clusters are significant because they offer a likely origin for the otherwise mysterious globular clusters. Despite their powerful gravity, they rapidly shed their star-forming gas, choking off further star formation after an initial burst. Short-lived monster stars created in that first wave age and die in just a few million years, generating huge supernova shock waves that soon blow away the surrounding nebula. Once the intermediate stars have also reached the end of their lives, all that remains is a compact globular cluster packed with many thousands of low-mass stars of identical age.

c.1970	1977	1978	2002
The Toomre brothers link computer models of galaxy collisions to peculiar galaxies	Alar Toomre suggests that merging spiral galaxies coalesce into ellipticals	Leonard Searle and Robert Zinn suggest that spiral galaxies form from the merger of smaller irregulars	Matthias Steinmetz and Julio Navarro use advanced computer models to back the theory of hierarchical galaxy evolution

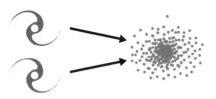

Left: In a major galaxy merger, colliding spirals lose their structure and coalesce to form a larger elliptical galaxy.

Right: In a minor galaxy merger, absorption of a small dwarf galaxy into a spiral enhances the spiral's structure and rate of star formation.

temperature to millions of degrees and enriching it with the products of their nuclear fusion. Eventually, the gas may be so hot and fast moving that it escapes into a halo region around the visible galaxy.

Despite major advances in both computing technology and our understanding of galaxy composition since those first simulations (including the discovery of dark matter – see Chapter 45), the Toomre model of 'major mergers' between large spiral galaxies has survived more or less intact. Of course, not all mergers involve a pair of spirals: encounters vetwee with smaller dwarf elliptical or irregular galaxies are far more common. These are much more one-sided affairs, in which the smaller galaxy is torn apart under the influence of the larger system, eventually losing its identity completely as it is cannibalized. As a side effect, the gravitational pull of the smaller galaxy seems to intensify the rate of star formation and the visible spiral pattern (see page 138). There is good evidence that our own galaxy is currently involved in such an episode at the moment, interacting with a small galaxy known as the Sagittarius Dwarf Elliptical.

COLLISIONS AS EVOLUTION

Based on his studies of the way in which the surviving stars from a major merger would behave, in 1977 Alar Toomre put forward the daring suggestion that mergers between spirals produce elliptical galaxies. The initial merger would disrupt the orbits of stars onto a variety of haywire elliptical paths, and the loss of much of the gas in the merged system removes a key influence that flattens out their orbits. Gas clouds naturally flatten into a disc as they collide, exerting gravitational pull on existing stars and controlling the plane in which new generations form. As the brighter and more massive stars created in the actual merger rapidly age and die, the ultimate result is an amorphous ball of more sedate red and yellow stars in overlapping orbits: an elliptical galaxy. Assuming that all galaxies started out as spirals, Toomre even calculated the likely rate of mergers over the lifetime of the Universe, showing that it matched the present proportion of elliptical galaxies.

The Toomres' ideas took some time to catch on and were fiercely debated through the 1980s, but more detailed observations of galaxy mergers gradually revealed many systems that seem to capture different phases in the transition from spiral to elliptical. Recent advances, meanwhile, have focused on filling in the gaps around the basic merger idea. By imaging galaxies many billions of light years away in an earlier epoch of cosmic evolution (see page 177), the Hubble Space Telescope has shown that most galaxies in fact started out as irregulars before merging and growing into more complex spirals. It's also become clear that merged ellipticals can gradually recapture gas from their surroundings. This allows them to regenerate through a lenticular phase (see page 146) and eventually form new spiral arms.

> **DOUBLE AND MULTIPLE SYSTEMS, AS WELL AS CLUSTERS, MAY BE EXPLAINED AS A RESULT OF CAPTURES BETWEEN NEBULAE, EFFECTED BY TIDAL FORCES AT CLOSE ENCOUNTERS.**
>
> Erik Holmberg

The entire merger cycle probably repeats several times, with the gas elevated to increasingly high temperatures and recaptured ever more slowly, in the course of a galaxy's evolution from a young spiral to a huge and ancient giant elliptical (see page 157) in the heart of a galaxy cluster.

The condensed idea
Galaxies collide frequently, and change their form as a result

38 Quasars and active galaxies

Active galaxies come in various forms, but are united by the presence of a bright and variable central nucleus, where a supermassive black hole is feeding on material from its surroundings. The best known of these galaxies are undoubtedly quasars, which have a key role to play in the story of galaxy evolution.

In 1908, astronomers Vesto Slipher and Edward A. Faith of California's Lick Observatory published details of strange features in the spectrum of Messier 77, one of the brightest galaxies in the sky. It stood out because its spectrum showed not the usual mix of absorption lines created by the light of countless stars, but emission lines – specific wavelengths of light that were so bright they stood out even against the 'continuum' of stellar spectra. Slipher and Faith did not know it at the time, but they had discovered the first active galaxy.

SEYFERT AND RADIO GALAXIES

It was not until 1943, when Carl Seyfert announced his discovery of a number of spiral galaxies with particularly bright, starlike points of light in their central nuclei, that galaxies with similar properties to M77 were found. The width of the emission lines indicated that they were produced by gas clouds orbiting the central region at high speed (causing their light emissions to be Doppler-shifted across a range of wavelengths – see page

TIMELINE

1943	1953	1960	1963
Seyfert identifies a number of spiral galaxies with bright compact nuclei and broad emission lines	Baade and Minkowski link the Cygnus A radio source to a distant peculiar galaxy	Sandage identifies the first quasi-stellar radio sources or quasars	Schmidt discovers the great distance of quasar 3C 273

62). Today, such systems are known as Seyfert galaxies, and they are recognized as the weakest form of active galaxy.

Meanwhile, in 1939, a young astronomer called Grote Reber identified some of the first astronomical radio sources aside from the Milky Way itself (see page 141). Pinning down visible objects corresponding to these radio sources proved difficult, however, since the resolution of early radio maps was severely limited. It was not until 1953, that Walter Baade and Rudolph Minkowski used more accurate radio surveys to locate Reber's sources. While most could be associated with objects in our own galaxy, such as supernova remnants, one, designated Cygnus A, appeared to be linked to a distant pair of colliding galaxies. A few months later, it became clear that the Cygnus A radio source actually consisted of two extended lobes, one on either side of the central galaxy system.

> THIS NUCLEAR REGION WOULD BE ABOUT 100 TIMES BRIGHTER OPTICALLY THAN THE LUMINOUS GALAXIES... IDENTIFIED WITH RADIO SOURCES THUS FAR.
> Maarten Schmidt

THE QUASAR MYSTERY

The late 1950s saw a flourishing in radio astronomy, with the development of the first large dish-shaped radio telescope at Jodrell Bank near Manchester, England. Many new extragalactic radio sources were discovered, and while some conformed to the double-lobed template of Cygnus A, many others consisted of just single sources. In 1960, US astronomer Allan Sandage led an effort to survey the sky around these objects, and found that they were typically associated with faint, starlike points of light. Sandage called them quasi-stellar radio sources, but within a few years this had been shortened to the rather more elegant 'quasar'. Their visible-light spectra seemed to show broad and bright emission lines far more powerful than those of Seyfert galaxies, but frustratingly, these could not be matched to any known elements.

A breakthrough in understanding finally came in 1963, when Sandage's Dutch colleague Maarten Schmidt realized that the spectral lines of a quasar

1964	1968	1969
Edwin Salpeter and Yakov Zel'dovich suggest that quasar emissions may come from the accretion disc around a giant black hole	John L. Schmitt discovers another type of active galaxy: the blazar or BL Lac object	Lynden-Bell argues that all active galaxies can be explained by the presence of a supermassive black hole

Mergers and active galaxies

Since the initial discovery of active galaxies, it's become clear that violent activity in the hub is frequently associated with the spectacular process of a galaxy collision or close encounter. For example, Centaurus A, one of the closest radio galaxies to Earth, appears in visible light as an elliptical galaxy called NGC 5128, crossed by a dark lane of opaque dust that is itself studded with areas of star formation and bright young clusters. The system is thought to be the result of a merger between an existing elliptical galaxy and a large spiral that has been effectively swallowed up. Such events inevitably result in large amounts of interstellar gas and even whole stars being driven into the reach of the central black hole, sparking it into life. Comparatively low-level activity, such as that seen in Seyfert galaxies, meanwhile, may be driven by tidal disruption from smaller galaxies merging with the larger system, or simply in orbit around it. Ultimately, once a galaxy merger process is complete, the individual supermassive black holes of the formerly independent systems may also spiral together and merge, generating powerful gravitational waves (see page 191) in the process.

called 3C 273, actually matched with the familiar emission lines produced by hydrogen, if they were shifted to the red end of the spectrum to an unprecedented degree. If, as seemed likely, this red shift was caused by the Doppler effect, it suggested that 3C 273 was moving away from Earth at one-sixth the speed of light.

Some astronomers attempted to explain this mysterious object as a runaway star, boosted to extreme speed by some hitherto unknown mechanism, but these efforts faltered as extreme red shifts were found in other quasars, but no similarly extreme blue shifts came to light (as might be expected if a random mechanism was operating). Instead, most experts concluded that quasars owed their high speed to the expansion of the Universe as a whole (see Chapter 40), and therefore, according to Hubble's Law, they must be both extremely distant and extremely bright. What was more, the light source involved must be comparatively tiny: the speed of unpredictable variations in quasar brightness showed they must be light hours across at most, and perhaps no larger than our solar system. Eventually, the theory that quasars were regions of intense activity embedded in the nuclei of distant galaxies was confirmed by observation of the much fainter 'host galaxies' surrounding them.

A UNIFIED THEORY

The links between these three types of active galaxy – radio-quiet Seyferts, radio galaxies and quasars – became clearer in the 1960s and 1970s. As the resolution of radio telescopes improved, it became clear that the lobes of radio galaxies were created as tightly focused jets of material emerging at high speed from the heart of the central galaxy encountered surrounding gas in the 'intergalactic medium' (see page 157) and billowed out into huge clouds. Some quasars also proved to have twin lobes of radio emission, while some Seyfert galaxies also emitted weak radio signals. The discovery of a new class of active galaxies called blazars added to the variety of activity observed.

As early as 1969, Donald Lynden-Bell argued that the behaviour of nearby radio and Seyfert galaxies could be a scaled-down version of quasar activity, and that all active galaxies ultimately owed their behaviour to a central giant black hole pulling in vast amounts of material from its surroundings. Though Lynden-Bell's idea was controversial at the time, growing evidence in its favour led to a unified model for Active Galactic Nuclei (AGNs) being developed in the 1980s. In this, radiation is emitted by an intensely hot accretion disc surrounding the central black hole, while jets of particles escaping from above and below the disc create the radio lobes. The exact type of galaxy we observe depends on the strength of activity and the AGN's orientation relative to Earth.

Jets billow out into radio lobes as they encounter intergalactic gas

Jets ejected along black hole's axis of rotation

Accretion disc generates intense radiation

AGN appears as quasar if accretion disc can be clearly seen

AGN appears as radio galaxy if central regions are blocked from view

Opaque dust ring blocks view of disc when seen edge-on

Central supermassive black hole

Weaker activity in AGN creates a Seyfert galaxy.

View straight down AGN jets gives rise to a 'blazar' galaxy

The complex structure of an AGN gives rise to various different types of active galaxy depending on the angle from which it is viewed.

The condensed idea
Monster black holes can create violent activity within galaxies

39 The large-scale Universe

Galaxies gather together on a variety of scales, with relatively compact groups and clusters overlapping at the edges to produce larger superclusters, and a cosmic-scale structure of filaments and voids. The distribution of different galaxy types not only reveals the secrets of galaxy evolution, but also tells us something important about conditions in the early Universe.

As astronomers discovered spiral and elliptical galaxies in growing numbers through the 18th and 19th centuries, their uneven distribution in the sky became obvious. The clearest cluster lay in the constellation of Virgo, but there were also prominent ones in Coma Berenices, Perseus, and the southern constellations of Fornax and Norma. Edwin Hubble's discovery of a relationship between a galaxy's distance and the red shift of its light in 1929 (see page 161) confirmed that these regions really did contain hundreds of bright galaxies crowded into a relatively small volume of space. Our own small 'Local Group' is far less impressive than these distant clusters. Identified by Hubble in 1936, this collection of a few dozen galaxies contains just three spirals – the Milky Way, Andromeda and Triangulum systems – and two large irregulars (the Magellanic Clouds).

Throughout the 1930s, many more galaxy clusters were identified, and astronomers began to apply a more sophisticated approach to analysing

TIMELINE

1929	1933	1936	1953
Hubble establishes the link between a galaxy's distance and the red shift of its light	Zwicky applies the Virial theorem to the Coma galaxy cluster and discovers dark matter	Hubble identifies the Local Group of galaxies close to the Milky Way	De Vaucouleurs suggests the existence of a supercluster incorporating the Local Group and Virgo Cluster

cluster membership, with clusters defined not by simple proximity, but by galaxies being held together by a gravitational attraction from which they cannot escape. Galaxy clusters are generally accepted to be the largest 'gravitationally bound' structures in the Universe. Because gravity diminishes rapidly with distance, clusters and groups typically occupy a space about 10 million light years across, regardless of how many galaxies they contain.

CHARACTERISTICS OF CLUSTERS

In 1933, Fritz Zwicky used a mathematical technique called the Virial theorem to estimate the Coma Cluster's mass from the velocities of its galaxies. This led him to predict that clusters contained vastly more matter and mass than their visible galaxies suggested. The first X-ray satellites launched in the 1970s, revealed that the centres of dense clusters were often sources of intense radiation, now understood to emanate from sparse 'intracluster' gas with temperatures of more than 10 million°C (18 million°F). This X-ray-emitting gas adds considerably to the mass of a cluster, but it still leaves the vast majority of Zwicky's missing material unaccounted for (see page 180).

Giant elliptical galaxies

Messier 87 at the centre of the Virgo Cluster is the largest galaxy in our region of the Universe. This huge ball of stars 120,000 light years across contains roughly 2.5 trillion solar masses of material. It is the archetypal giant elliptical galaxy, also known as 'central dominant' or 'type cD' galaxy. Giant ellipticals are galactic cannibals, the end result of multiple galactic mergers that have seen smaller ellipticals and spirals swallowed up. As a result, they are often surrounded by faint haloes of stars that extend to a total diameter of perhaps half a million light years. These are the stray survivors from past collisions, flung off into wild orbits around the central galaxy. They also frequently have retinues of globular clusters orbiting in the same region – Messier 87 has around 12,000 (compared to the Milky Way's 150 or so) and, if the link between globulars and super star clusters (see page 149) is correct, this is also likely to be due to cosmic collisions. Further evidence to support the growth of giant ellipticals from galaxy collisions comes from several examples that conceal more than one supermassive black hole in their cores. Messier 87 has just one, but thanks to its most recent merger, it is an active galactic nucleus and one of the brightest radio sources in the sky.

1958	1977	1982–85	2014
Abell publishes the first version of his catalogue of galaxy clusters	Astronomers at the Harvard-Smithsonian Center for Astrophysics begin the first large-scale galaxy red shift survey	Results from red shift surveys reveal the cosmic structure of filaments and voids	The Virgo Supercluster is replaced by the larger structure called Laniakea

Another important aspect is clusters' distinct mix of galaxy types. 'Field galaxies' – the 20 per cent of nearby galaxies unattached to any particular cluster – are typically irregular or spiral, while galaxies in loose groups like our own come in all forms. Dense clusters, however, are dominated by ellipticals, and the very centre is often marked by a truly enormous giant elliptical (see box on page 157). In 1950, Lyman Spitzer, Jr. and Walter Baade argued that this distribution indicates that ellipticals evolve through collisions, which are more likely to happen in the crowded environments of dense clusters. They even predicted that such collisions would strip galaxies of interstellar gas, a result that pre-empted theories of galaxy evolution in the 1970s (see page 149), and which was borne out by the discovery of the X-ray-emitting intracluster gas.

> WE HAVE FINALLY ESTABLISHED THE CONTOURS THAT DEFINE THE SUPERCLUSTER OF GALAXIES WE CAN CALL HOME.
> R. Brent Tully,
> on the Laniakea Supercluster

In the 1950s, George Ogden Abell began compiling an exhaustive catalogue of clusters, which would not be completed until 1989. Abell's catalogue led to many important new discoveries, but perhaps the most notable was the 'cluster luminosity function' – the relationship between the intrinsic brightness of a cluster's most luminous galaxy, and the numbers of galaxies above a particular brightness level. Since the *relative* brightness of a cluster's galaxies is easily measured, the luminosity function provides a way of predicting their true luminosities, and is therefore an important 'standard candle' for measuring large-scale cosmic distances.

STRUCTURE BEYOND CLUSTERS

Both Abell and French–American astronomer Gérard de Vaucouleurs argued for the existence of a further level of structure beyond galaxy clusters. In 1953, De Vaucouleurs suggested the existence of a 'Local Supergalaxy' centred on the Virgo Cluster and encompassing many other clusters including our own Local Group, but it was not until the early 1980s that red-shift surveys proved its existence beyond doubt. Several other 'superclusters' were soon identified, but their precise definition remains open to debate, since they are not gravitationally bound in the same way as galaxies in individual clusters. Instead, superclusters are simply defined as concentrations of clusters in a region of space, often with a general shared motion. This fluid definition is one reason why, in 2014, the Virgo Supercluster was swept aside in favour of a new and larger structure, called Laniakea, some 500 million light years across and containing at least 100,000 major galaxies.

Advances in technology since the 1970s have enabled the collection of spectra and red shifts for huge numbers of galaxies, allowing accurate maps of the large-scale Universe to be created. These show that superclusters merge at their edges to form chainlike filaments, hundreds of millions of light years long, around vast and apparently empty regions known as voids. This unexpected discovery ran counter to assumptions that the cosmos would be essentially the same in all directions. While uniformity does seem to set in across even larger scales of billions of light years, the largest structure we do see cannot possibly have been created by gravitational interactions over the lifetime of the cosmos. This sets important restrictions on how the Universe itself formed (see Chapter 41).

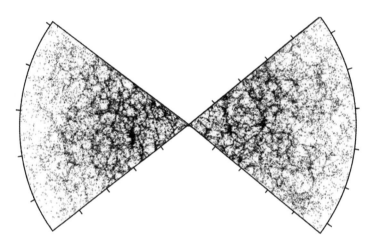

A slice from the Anglo-Australian Observatory's 2dF (Two-degree Field) Galaxy Redshift Survey reveals the distribution of tens of thousand of galaxies in a cosmic web of filaments and voids.

The condensed idea
Structure is found at all scales of the cosmos

40 The expanding cosmos

The surprising finding that the Universe as a whole is expanding revolutionized astronomy in the mid-20th century, although astronomers argued about its meaning for several decades. Meanwhile, the true rate of expansion, with important implications for the origin and fate of the Universe, was not established until surprisingly recently.

Credit for discovering the expanding Universe is usually given to Edwin Hubble, who made the pioneering galaxy distance measurements and key calculations around 1929, but the real story is more complex. In 1912, Vesto Slipher at the Lowell Observatory in Flagstaff, Arizona, surveyed the spectra of spiral nebulae and found that they mostly had large red shifts. Assuming it to be a Doppler effect due to the nebulae moving away from us, Slipher calculated that they were receding at speeds of hundreds of kilometres per second. This was important evidence that the nebulae were not simply small star clouds in orbit around the Milky Way, but the clinching proof came from Hubble's measurements of Cepheid variables (see page 146).

THEORY AND PRACTICE
In 1915, Albert Einstein published his theory of general relativity (see page 190). It passed early tests with flying colours, but it created a major problem for theories of the Universe: according to the simplest interpretation, the presence of large amounts of mass within the Universe would inevitably

TIMELINE

1912	1922	1927	1929
Slipher discovers the large red shift of many spiral nebulae	Friedmann finds a solution to general relativity in which the Universe is expanding	Lemaître predicts that more distant galaxies should show larger red shifts	Hubble identifies the relationship between galaxy red shift and distance

lead it to collapse in on itself. The scientific consensus at the time was that the Universe was infinitely old and static, so Einstein resolved the problem by adding a 'cosmological constant' to his equations. This weak antigravity force acted only on the largest scales to counteract the contraction of space. He would later call this his greatest mistake, although the more recent discovery of dark energy has somewhat vindicated the idea (see page 186).

In 1922, Russian physicist Alexander Friedmann came up with an alternative solution to Einstein's equations, showing that they were equally valid if spacetime was expanding, but his work was largely ignored because of a lack of evidence to support it. A few years later, in 1927, Belgian astronomer and priest Georges Lemaître reached similar conclusions, but crucially he predicted an observational consequence: all galaxies should be moving away from each other on the largest scales, and the further away a galaxy is, the greater the speed of its recession from the Milky Way.

Neither Friedmann nor Lemaître's work seems to have directly influenced Hubble when he set out in the late 1920s to compare his measurements of galaxy distances with the red shifts recorded by both Slipher and Hubble's colleague Milton Humason. However, he quickly discovered the exact relationship predicted by Lemaître, and in 1929 published his evidence, including a graph showing the link between galaxy velocity and distance. This relationship is now known as Hubble's Law, while the graph's gradient – the rate at which the speed of galaxy recession increases with distance – is called the Hubble Constant (denoted H_o).

> **THEORIES CRUMBLE, BUT GOOD OBSERVATIONS NEVER FADE.**
> Harlow Shapley

THE MEANING OF EXPANSION

Hubble's discovery had huge implications for the history of the Universe, even if Hubble himself was slow to embrace them (see box on page 162). If everything in the Universe is moving away from the Milky Way, then the two possibilities

1931	1958	2000
Lemaître argues that cosmic expansion indicates the Universe began in a high-temperature primeval atom	Sandage delivers the first modern estimate of the Hubble Constant	Publication of results from the Hubble Key Project

Hubble's error

Edwin Hubble's discovery of the link between red shift and galaxy distance was hugely important, but Hubble himself ultimately rejected the idea of an expanding Universe. At the time of his measurements, astronomers did not fully distinguish between 'classical Cepheids' (Population I stars containing substantial amounts of heavy elements) and a slightly fainter group of Population II stars (with their own distinct relationship between period and luminosity). This led Hubble to significantly underestimate intergalactic distances (for example, he placed the Andromeda Galaxy 900,000 light years away, when modern measurements suggest a distance of 2.5 million light years).

As a consequence of this, Hubble also overestimated the rate at which red shift increased with distance, and therefore calculated that if red shifts were due to the Doppler effect, then recession must be increasing with distance at 500 km/s/Mpc. Tracing this expansion backwards in time would see all galaxies coincide at the same point in space (Lemaître's primeval atom) just 2 billion years ago. Since this was less than half the age of the Earth as known at the time, Hubble dismissed the Doppler explanation in favour of other hypothetical ideas, such as the concept of 'tired' light that becomes increasingly red shifted over great distances due to other effects.

are that our region of space is so uniquely unpopular that galaxies really are fleeing away from it, or that the Universe as a whole is expanding, and all the galaxies within it are being carried away from each other as Lemaître predicted. No astronomer took the first option seriously, since it implies that we have a privileged position in the Universe (counter to many harsh lessons learned since the time of Copernicus). But the suggestion that an expanding Universe carried with it, of an origin in the measurable past, was almost as uncomfortable for astronomers who generally believed in an eternal cosmos. It was Lemaître who fully embraced them in 1931, arguing that cosmic expansion implies a Universe that was hotter and denser in the past, ultimately originating in a 'primeval atom'. This was the precursor of the modern Big Bang theory (see page 164).

With the 1964 discovery of the Cosmic Microwave Background Radiation (see page 178), most cosmologists regarded the case for the Big Bang as proven. Accurate measurement of the Hubble Constant now took on a new importance, since the rate of present-day expansion can be turned on its head to estimate the age of the Universe. As observing technology improved, so did the ability to detect Cepheid variable stars in more distant galaxies (and to distinguish between Cepheids and the misleadingly similar RR Lyrae stars). In 1958, Allan Sandage published a vastly improved estimate of H_o, suggesting that the rate of recession for distant galaxies increased by 75 kilometres per second for each megaparsec of distance (Mpc, a unit equivalent

A common analogy when considering the expansion of the Universe is to imagine space as an inflating balloon. As the balloon stretches, points on its surface (galaxies) move away from each other. The greater their initial separation, the faster they move apart. The stretching of light waves by expanding space can also be described in a similar way.

to 3.26 million light years). This was about one sixth of Hubble's value (see box, left), and implied that the Universe was a far more plausible 13 billion years old.

Over the next few decades, measurements of H_o fluctuated significantly around Sandage's value, between around 50 and 100 km/s/Mpc, implying a Universe between 10 and 20 billion years old. Settling the matter was the Hubble Space Telescope's 'Key Project', and drove the telescope's design from its inception in the 1970s through to its eventual launch in 1990. In between more high-profile observations, the HST spent much of its first decade gathering data and measuring Cepheid light curves in galaxies out to about 100 million years, leading to final publication in 2000 of a value of 72 km/s/Mpc. Further measurements have hovered around the same value, resulting in a widely agreed age for the Universe of 13.8 billion years.

The condensed idea
The Universe is getting bigger with every passing moment

41 The Big Bang

The idea that the Universe began in a huge explosion some 13.8 billion years ago is the mainstay of modern cosmology, and is key to explaining many observed aspects of the Universe. Yet when it was first put forward, the idea of a finite universe was anathema to many in the scientific establishment.

While Russian physicist Alexander Friedmann showed as early as 1922 that an expanding Universe was consistent with Einstein's theory of general relativity (see page 161), credit for the Big Bang is usually given to Belgian priest Georges Lemaître, who published his 'primeval atom' theory in 1931. At first glance it might seem strange for a Catholic priest to have made such a fundamental contribution to modern physics, but Lemaître had studied cosmology at Cambridge under Arthur Eddington, and at Harvard with Harlow Shapley. He had also argued for cosmic expansion well before it was confirmed by Edwin Hubble.

RIVAL THEORIES

For three decades, Lemaître's theory was viewed as just one of several competing explanations for cosmic expansion. Unable to accept the concept of a moment of creation, Friedmann argued for a cyclic Universe that went through phases of alternating expansion and contraction. In the 1940s, meanwhile, Hermann Bondi, Thomas Gold and Fred Hoyle published arguments in favour of a 'steady state' Universe – a perpetually expanding cosmos in which matter was continuously created to maintain a constant density. In 1948, physicists Ralph Alpher and Robert Herman predicted that

TIMELINE

1931	1948	1948
Lemaître argues from cosmic expansion that the Universe originated in a hot, dense primeval atom	Alpher and Gamow show how conditions in the early Universe could give rise to elements	Alpher and Hermann predict the existence of radiation from the edge of space as a consequence of the primeval atom theory

Lemaître's primeval fireball would have left a detectable afterglow, equivalent to the radiation from a black body a few degrees above absolute zero. This Cosmic Microwave Background Radiation could not plausibly be created by any of the rival theories, and its somewhat accidental discovery by Arno Penzias and Robert Wilson in 1964 was crucial evidence for a 'Big Bang' moment of creation (see page 178).

> **YOUR CALCULATIONS ARE CORRECT, BUT YOUR PHYSICS IS ATROCIOUS.**
> Albert Einstein to Georges Lemaître

The challenge for any creation theory is to produce a Universe with conditions similar to those we see today, and here the Big Bang proved its merit well before the Penzias and Wilson discovery. Key evidence lay in the fact that mass and energy are equivalent and can be exchanged in extreme situations, a fact encapsulated in Einstein's famous equation $E = mc^2$ (see page 189). Thus, if cosmic expansion was traced back to the earliest times, rising temperatures would see matter disintegrate into its component particles, and ultimately disappear in a blizzard of pure energy. In 1948, Ralph Alpher and George Gamow published a landmark paper showing how the falloff from this intense fireball would produce elements in identical proportions to those expected in the early Universe (see page 169).

THE STRUCTURE PROBLEM

While subsequent theoretical work, as well as results from early particle accelerators (see box, page 166), backed up the idea of raw cosmic elements being forged in the Big Bang, discoveries about the structure of the Universe in the 1970s raised new questions. While extremely complex, they boiled down to the essential puzzle of how the Big Bang could produce a cosmos smooth enough to not show vast differences from place to place (as reflected by the apparently uniform temperature of the CMBR), yet somehow varied enough to give rise to the large-scale structure of superclusters, filaments and voids (see page 159). The basic Big Bang theory predicted a primeval fireball in which matter was smoothly distributed. Until about 380,000 years after the Big Bang, incandescent

1949	1964	1981	1992
Hoyle coins the term 'Big Bang' as an insult to the theory	Penzias and Wilson discover the Cosmic Microwave Background Radiation	Alan Guth proposes inflation as a means of producing the observed structure in the Universe	The COBE satellite maps the CMBR, confirming the presence of structure in the Universe at the very earliest times

From energy to matter

Much of our understanding of the Big Bang, and particularly of the way that raw energy rapidly gave rise to matter, is derived from experiments using particle accelerators. These enormous machines use powerful electromagnets to boost charged subatomic particles to near light-speed, then slam them together and monitor the results. Collisions such as those at the Large Hadron Collider in Switzerland transform small amounts of matter into pure energy, which then condenses back into a shower of particles with different masses and properties. In this way, we know that relatively heavy particles called quarks were only able to form in the incandescent temperatures of the first millionth of a second after the Big Bang itself, after which they rapidly bound together in triplets to form the protons and neutrons needed for nucleosynthesis (see page 168). Lighter lepton particles (principally electrons) continued forming until the Universe was about 10 seconds old.

Intriguingly, however, there's nothing inherent in the Big Bang to explain why today's Universe is dominated by familiar 'baryonic' matter particles rather than antimatter (mirror-image particles with opposite electric charges). In fact, most cosmologists believe that the initial explosion created equal amounts of matter and antimatter particles, the vast majority of which collided to annihilate each other in a burst of energy. Some unknown 'baryogenesis' process ensured there was a tiny excess of normal matter left over at the end, and this is what accounts for all the baryonic matter in the Universe today.

temperatures prevented atomic nuclei from joining with electrons to make atoms, and the high-particle density repeatedly deflected and scattered photons of light, preventing them from travelling in straight lines (more or less the same thing that happens in thick fog). In this environment, radiation pressure on the particles would overcome gravity and prevent them from coalescing into the seeds of structure needed to give rise to today's huge filaments. Eventually, the cosmic temperature cooled enough for electrons and nuclei to join up, particle density fell and the fog suddenly cleared. The light escaping this event – known as the 'decoupling' of radiation and matter – now forms the CMBR. By this time, matter would be dispersed too widely to form supercluster structures, and perhaps even spread too thinly to create galaxies.

Clearly, then, something else was going on. In 1981, Alan Guth at the Massachusetts Institute of Technology proposed a possible solution: what if some cataclysmic event in the first instants of the Big Bang had taken one tiny, essentially uniform fragment of the primeval Universe and blown it up to enormous size? The resulting bubble of space and time, encompassing the entirety of our observable Universe and far beyond, would show an effectively uniform temperature, but tiny variations, arising from the inherent uncertainties of

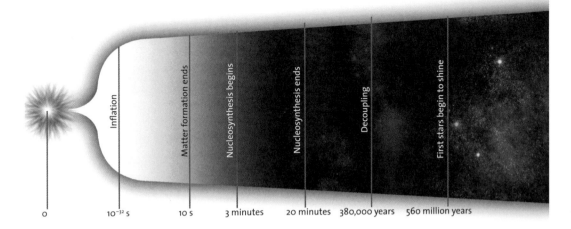

Inflation

Matter formation ends

Nucleosynthesis begins

Nucleosynthesis ends

Decoupling

First stars begin to shine

| 0 | 10^{-32} s | 10 s | 3 minutes | 20 minutes | 380,000 years | 560 million years |

A simplified timeline shows the major stages in the development of matter from the Big Bang itself, to the formation of the first stars and galaxies.

quantum physics, would be blown up to a vast scale, creating cool, relatively sparse patches alongside warmer, denser regions. Over time, minor variations could act as the nuclei around which matter accumulated.

Guth's theory, soon named inflation, was enthusiastically adopted by many others including Andrei Linde (see page 198). The plausibility of their models was helped by growing recognition of the role played by so-called dark matter, which would be immune to the radiation pressure forcing normal matter apart (see Chapter 45) and therefore able to begin the formation of early structure well before the decoupling stage. This idea was vindicated in spectacular style in 1992 by the results of the COBE satellite (see page 179) and has been further supported by other experiments. While cosmologists are still wrestling with some of inflation's wider implications, it forms a key element in the Big Bang as we understand it today.

The condensed idea
The Universe began in a hot, dense explosion of energy

42 Nucleosynthesis and cosmic evolution

How did the Big Bang give rise to the raw materials of the cosmos, and how did they subsequently change over time to create the mix of matter we see in the Universe today? The answers lie in a variety of different processes united under the name nucleosynthesis.

All matter in the Universe today is made of atoms, and every atom consists of an atomic nucleus (a clump of relatively heavy protons and neutrons) surrounded by a cloud of far lighter electrons. Atoms of different elements are distinguished from each other by the number of protons in the nucleus, while neutrons influence their stability. So the manufacture of elements is principally a matter of creating different nuclei in the process known as nucleosynthesis.

Establishing different chains of nucleosynthesis was a running theme of 20th-century astrophysics. For example, within low-mass main sequence stars the proton–proton chain and CNO cycle (see pages 74 and 75) both offered ways of transforming hydrogen nuclei (the simplest atomic nucleus, consisting of a single proton) into helium. The triple-alpha process (see page 110) in red giants, meanwhile, allows helium nuclei to build into carbon and oxygen, and nuclear fusion in supergiant stars goes much further to create increasingly complex elements up to iron and nickel (see page 118). Finally,

TIMELINE

1904	1930	1948
Hartmann identifies the existence of cold interstellar gas through its effect on stellar spectra	Robert Trumpler demonstrates the absorption effects of the Milky Way's interstellar dust	The Alpher Bethe Gamow paper outlines the way in which elements can be formed in the Big Bang

supernova explosions provide the final rung on the ladder towards the heaviest natural elements (see page 122).

BUILDING THE FIRST ATOMS
But how did hydrogen itself, the first rung on that ladder, come into being? The essentials were worked out in the late 1940s by George Gamow and Ralph Alpher in a theory generally known as Big Bang nucleosynthesis. The pair built on Gamow's previous work to imagine a rapidly expanding primeval fireball in the early Universe, composed entirely of neutrons that begin to spontaneously decay into protons and electrons as the surrounding pressure falls. The

Alpher, Bethe and Gamow

The short 1948 paper that first outlined Big Bang nucleosynthesis had not two, but three authors – Ralph Alpher, Hans Bethe and George Gamow. Gamow whimsically included a credit to his colleague Bethe in absentia as a play on the first three letters of the Greek alphabet (alpha, beta and gamma). Alpher, as a graduate student working on his PhD at the time, admitted to being less than impressed by Gamow's little joke, fearing that his contribution would be overshadowed by sharing with not one but two highly respected astrophysicists. Nevertheless, Bethe did help out by reviewing the paper prior to publication, and subsequently contributed to the further development of the theory.

formation of nuclei more complex than hydrogen therefore becomes a race against time – how many neutrons can be swept up by protons to form heavier nuclei, *before* the neutrons themselves decay?

When Alpher and Gamow looked at the problem in terms of the possibilities for various particles to capture neutrons, they found that the most abundant elements in the Universe by far would be hydrogen, accounting for 75 per cent of cosmic mass, and helium, accounting for the remaining 25 per cent. Tiny amounts of lithium and beryllium would also have been formed in this way, and these predictions subsequently turned out to match new measurements of the cosmic abundance of elements. The only significant mistake in the paper was the authors' assumption that *all* the elements would have to be created by neutron capture in this way. In

1952
Fred Hoyle and Alfred Fowler discover the triple-alpha helium fusion process for building elements such as carbon

1957
The B2FH paper demonstrates how heavy elements are formed in the most massive stars and supernovae

1961
Guido Münch and Harold Zirin find evidence for gas clouds in the galactic halo, and a hot galactic corona

1977
Christopher McKee and Jeremiah Ostriker put forward a three-phase model for the intergalactic medium

reality, this is impossible because of 'mass gaps' where nuclei with certain configurations disintegrate as swiftly as they can form. Such gaps cannot be bridged by adding particles one step at a time, and mean beryllium is the heaviest element that can be created from scratch by this route. Instead, manufacturing heavier elements requires a leap in particle numbers of the kind that only the triple-alpha process can supply.

STARSTUFF

A better understanding of how elements are formed and how their abundances change over time has given rise to a far more cyclical view of stellar life cycles. At the same time, it has revealed a deeper picture of the relationship between stars and the interstellar medium (ISM), the material that surrounds them and from which they are born.

> **WE ARE WE ARE BITS OF STELLAR MATTER THAT GOT COLD BY ACCIDENT, BITS OF A STAR GONE WRONG.**
> Arthur Eddington

The evidence for large clouds of material between the stars was discovered in the first half of the 20th century. E.E. Barnard gets much of the credit for his photography of dark nebulae – opaque clouds of gas and dust that are only visible when silhouetted against a brighter background – but German astronomer Johannes Hartmann was the first to prove the existence of cold, invisible gas clouds by recognizing the faint imprints their absorption lines left on the spectra of more distance stars (see page 60).

Since the 1970s, most astronomers have subscribed to a three-phase model of the ISM, with the different phases distinguished from each other by their temperature and density. The cold phase consists of relatively dense clouds of neutral hydrogen atoms at just a few tens of degrees above absolute zero; the warm phase contains much hotter neutral and ionized hydrogen with temperatures of thousands of degrees; and the hot phase consists of highly dispersed ionized hydrogen and heavier elements with temperatures of a million degrees or more.

In the so-called 'galactic fountain' model of cyclical evolution, ISM material sits out in the cold dense phase before some external influence (perhaps an encounter with a passing star, passage through a spiral density wave or the shockwave from a nearby supernova) encourages it to collapse under its own gravity, beginning the process of star formation (see Chapter 21). Once the first stars emerge within the medium, their radiation warms and ionizes the surrounding gas, creating a glowing starbirth nebula. As the most massive of the newborn stars hurtle towards the end of their lives, strong stellar winds and supernova shockwaves create huge bubbles in the ISM, with some material heated so much that it escapes from the galactic disc entirely to form so-called 'coronal gas'. Over millions of years, this hot ISM gradually cools and sinks back towards the disc, enriching it with further heavy elements.

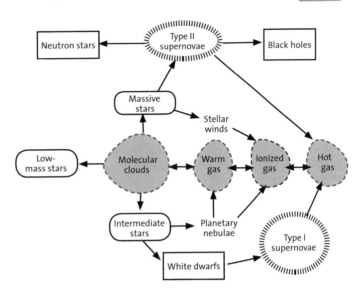

This schematic shows key elements of the 'galactic ecology' through which matter is processed in stars and returned to the interstellar medium.

This is just a broad picture of the process at work in a typical galaxy, but as the same events are repeated throughout the cosmos, they will gradually enrich it with increasing amounts of the heavier elements. Nevertheless, it seems unlikely that stars will run out of fuel any time soon – gas in our galaxy's ISM today is still 70 per cent hydrogen and 28 per cent helium by mass, with only 1.5 per cent heavier elements to show for more than 13 billion years of stellar nucleosynthesis.

The condensed idea
Our Universe is a factory for manufacturing elements

43 Monster stars and primordial galaxies

The earliest objects in the Universe are currently beyond the range of even the most advanced telescopes. Nevertheless, most astronomers believe the evidence points to an initial generation of short-lived but spectacular giant stars whose violent deaths created conditions for the formation of later galaxies.

One of the central questions in cosmology is whether the large-scale structure of the Universe formed 'bottom up' or 'top down'. In other words, did small objects form more or less uniformly and then pull together through gravity to create larger structures, or did initial differences in the large-scale distribution of matter, seeded in the immediate aftermath of the Big Bang itself (see page 167), govern where matter coalesced?

TOP DOWN OR BOTTOM UP?

Current evidence suggests a mix of both processes at work. Large-scale differences in the distribution of matter are responsible for the general arrangement of galaxy superclusters into huge filaments around apparently empty voids. Meanwhile, smaller-scale structures, from galaxies up to clusters, are pulled together by the power of gravity.

This raises the question of what were the first smaller-scale structures that became the seeds of galaxies? The presence of supermassive black holes in

TIMELINE

1974	1978	2002
Cameron and Truran propose the existence of distinct Population III stars	Rees proposes Population III stars as a possible source of dark matter MACHOs	Bromm, Coppi and Larson show how the very first stars could overcome modern limits on stellar mass

the centre of most galaxies, and the dominance of brilliant quasars in the early Universe, suggests a sequence of events in which giant black holes pulled in material, gorged themselves to form quasars, and triggered waves of starbirth in the material that gathered around them at a safer distance. But where did those black holes come from in the first place?

Astronomers have pondered this scenario since the 1970s, mostly using computer models to show how matter could have collapsed and coalesced under the influence of gravity. While some have argued that supermassive black holes could form simply from the collapse of gas clouds in the early Universe, others suggest that they are more likely to have formed from the merger of smaller black holes left by the very first generation of stars.

The reionization problem

Intense radiation from Population III stars offers a potential solution to one of the biggest mysteries about the large-scale Universe: the so-called reionization problem. Put simply, the problem arises because the vast clouds of hydrogen found in intergalactic space are in an electrically charged or ionized form, with atoms stripped of their electrons. Yet according to the Big Bang theory, matter should have emerged from the primeval fireball in the form of uncharged atoms – indeed, the 'recombination' of atomic nuclei with electrons was the final stage of the Big Bang itself (see page 166). It seems that some process had to reionize the intergalactic medium before the first galaxies formed, and high-energy ultraviolet radiation from monster stars is thought to be the most likely candidate.

POPULATION III

Comparing stars with different ages at different locations in our galaxy and the wider Universe shows that the proportion of heavier elements, which astronomers call metals, within the raw materials of star formation has increased over the billions of years of cosmic history. Walter Baade recognized the difference between the young, metal-rich Population I and the older, metal-poor Population II in 1944, but it wasn't until the 1970s that the possibility of a distinctive Population III, made entirely from light

2003

Alexander Heger *et al.* model the processes that end the lives of the most massive stars

2005

Spitzer Space Telescope detects an infrared glow thought to originate from Population III stars elements formed in the Big Bang, was raised by A.G.W. Cameron and James Truran. The case for Population III stars became more pressing in the 1990s, after astronomers found that even the most distant and ancient quasars and primordial galaxies were already enriched with heavy elements from some earlier source.

> **THOSE STARS WERE THE ONES THAT FORMED THE FIRST HEAVY ATOMS THAT ULTIMATELY ALLOWED US TO BE HERE.**
>
> David Sobral

Around this time, cosmologists began to study the evolution of the early Universe through computer models. Starting with data from the irregularities of the Cosmic Microwave Background Radiation (see page 178), they traced how both luminous and unseen dark matter (see page 188) would behave. They found that within about 200 million years of the Big Bang, small 'protogalaxies' began to coalesce. With each containing up to a million solar masses of star-forming gas in a region a few tens of light years across, these would make ideal birthplaces for Population III stars.

MODELLING MONSTERS

Meanwhile, other astronomers were modelling the properties of the stars themselves. It soon became clear that because protogalaxy gas was much warmer and faster-moving than the present-day interstellar medium, much more gravity was needed to make it collapse into a star. In other words, the smallest initial star-forming clumps would be tens, perhaps hundreds of times more massive than those in today's Universe. In normal circumstances, this would be a recipe for disaster and disintegration – as the centre of the collapsing cloud heated up through gravitational collapse, it should pump out so much radiation that the outer regions would be blown away. But in 2002, researchers showed that the unique circumstances of the early Universe, with normal and dark matter still in close proximity and no heavy elements, could overcome this problem, allowing the formation of stars with many hundreds of solar masses.

Once formed, these monster stars would be surprisingly stable and subdued. Their lack of heavy elements would initially restrain their nuclear fusion processes to the simple proton–proton chain (see page 74), reducing the amount of radiation generated and preventing them from blowing apart.

Despite this, conditions in the core meant that these stars would still burn through their core hydrogen fuel supply in a few million years, and begin to create heavier elements in a similar way to today's red giants and supergiants. Eventually, though, these fuel sources too would be exhausted, and the stars would meet their end in spectacular supernovae far more powerful than any we know today (see box). In the process, they would scatter their heavy elements across surrounding space, enriching the mix of material in the larger galaxies that were already coalescing around them.

It's worth noting that the mass of these first stars is still a subject of debate. Some lines of evidence suggest that they were limited to masses more similar to those in the Universe today. Evidence of which model is correct may come from NASA's James Webb Space Telescope, which hopes to capture light from these Population III stars for the very first time after its launch some time around 2018.

The death of giants

Monster stars such as those that may have formed in Population III are thought to die in a unique type of supernova – a so-called photodisintegration hypernova. Photodisintegration is a process that takes place to some extent in the cores of all supernova-bound stars, and involves the fragmentation of atomic nuclei when they are struck by high-energy gamma rays. The process normally absorbs energy, and makes a small contribution to the overall process of nucleosynthesis in supernovae, but as stars of more than 250 solar masses approach the end of their lives, it can take place at a vastly accelerated rate. The absorption of energy creates a rapid pressure drop in the core of the star, producing a black hole that eats the star from the inside. A proportion of the star's material, enriched with the heavy elements created during its lifetime, may spray out from the poles in two jets at close to the speed of light, but the vast majority of the star's mass is retained by the black hole, making this a potential way to rapidly build black holes with hundreds of solar masses.

The condensed idea
The first protogalaxies were populated with monster stars

44 The edge of the Universe

Light might be the fastest thing there is, but its speed is still finite. This means that as we look across the vastness of space, we also look back in time. And because the Universe has a finite history, the limited speed of light also creates a cosmic boundary, beyond which we can never see.

The fact that light travels through space at about 300,000 kilometres per second (186,000 miles per second) was established in the 18th and 19th centuries. Results from a variety of ingenious experiments were backed up theoretically by the calculations of Scottish physicist James Clerk Maxwell, who showed in a landmark paper of 1864 that light is an electromagnetic wave – a combination of electrical and magnetic disturbances that propagate across space at a fixed speed.

The limited speed of light turns our Universe into a sort of cosmic time machine, since the light from distant objects must have taken some time to reach us. The very first plausible attempt to measure the speed of light, by Danish astronomer Ole Rømer in 1676, relied upon this very idea. Rømer noted changes in the timing of eclipses caused as the Galilean satellites of Jupiter (see page 8) moved around their parent planet, and attributed them to changes in the time light takes to reach Earth thanks to the varying positions of the two planets.

TIMELINE

1864	1948	1964
Maxwell establishes the fixed speed of light in a vacuum	Alpher and Hermann predict that the edge of the observable Universe should emit weak radiation	Penzias and Wilson detect radio signals from the Cosmic Microwave Background Radiation

In most situations, astronomers just take this effect, known as lookback time, for granted, but over the largest distances it has some useful side effects. When we look at objects billions of light years away in space, we are also seeing them billions of years back in history. Look far enough away, and the light from galaxies arriving in our telescopes left on its long journey towards Earth at a significantly earlier point in their evolution. This explains why violent active galaxies, such as quasars (see page 153), tend to be so far away in space – they represent a much earlier phase of galactic evolution in which supermassive black holes were feeding far more voraciously than they do in today's relatively quiet, evolved galaxies.

PROBING THE PAST

Since the 1990s, astronomers have used the unique capabilities of the Hubble Space Telescope (HST) to take advantage of this effect, creating a series of 'Hubble Deep Fields' that combine the faint light captured over many hours as the telescope stares unblinking at a single, apparently empty region of space. Several different areas of the sky have been studied in this way, and all reveal a similar story – countless galaxies stretching away to the limits of visibility. Elliptical galaxies only appear in the foreground of such images, while the middle ranges show spirals in the process of formation. The most distant galaxies are overwhelmingly irregular and illuminated by violent star formation.

The observable Universe

The ultimate limit on our observations of the Universe is set by how far light can have travelled in the estimated 13.8 billion years since the Big Bang (see page 164). This boundary, where the CMBR originates, is said to be the edge of the 'observable Universe'. One might reasonably assume that it lies 13.8 billion light years away in every direction. However, the reality is rather more complicated. The expansion of space as light has travelled across it has not only extended and reddened its wavelengths, but also increased the distance between its source and Earth. So while a beam of light itself may have travelled for 13.8 billion years, cosmic expansion means that its source is now far more than 13.8 billion light years away. In fact, the most recent estimates suggest that we can hypothetically see light from objects that are some 46.5 billion light years away, and therefore this is the true limit of the observable Universe.

1992
The COBE satellite measures ripples in the CMBR, the first hints of structure in the Universe

2005
J. Richard Gott III et al estimate the radius of the observable Universe at about 46.5 billion light years

2009
The European Space Agency's Planck satellite is launched, mapping the CMBR in unprecedented detail.

> **I HAVE OBSERVED STARS OF WHICH THE LIGHT, IT CAN BE PROVED, MUST TAKE TWO MILLION YEARS TO REACH THE EARTH.**
>
> William Herschel

Ultimately, however, the most distant galaxies suffer such huge red shifts that their light is mostly emitted in the infrared. The HST carries near-infrared instruments that allow it to follow galaxies a little way beyond the limits of visible light, but not far, and there is still a limit to the faintness of galaxies that even long-exposure Deep Field images can pick up. The most distant objects so far imaged are therefore rare galaxies whose (mostly infrared) light is amplified by the effect known as gravitational lensing (see page 190). However, it's generally thought that the incredibly powerful yet short-lived gamma ray bursts that occasionally reach Earth from all parts of the sky originate from cataclysmic events in galaxies that are otherwise currently undetectable (see pages 123 and 131).

NASA's James Webb Space Telescope, the infrared successor to Hubble, should be able to image many of these ancient galaxies and other objects in the early Universe (see page 175), but at present, the edges of the Universe finally fade into darkness around 13 billion light years away – a frustrating few hundred million years after the Big Bang itself. Fortunately, however, that's not quite the end of the story.

SIGNALS FROM THE EDGE

In 1964, radio astronomers Arno Penzias and Robert Wilson, working on a new and highly sensitive radio antenna at the Bell Telephone Laboratories in New Jersey, found their system plagued with an unknown source of faint but persistent radio noise. After investigating all possible sources of contamination (including the possibility of radio-emitting droppings from pigeons that nested in the antenna), they concluded that the signal was real. What was more, the radio noise was coming from all over the sky and corresponded to a uniform black body (see page 58) with a temperature of about 4 kelvin (4°C above absolute zero). This matched almost perfectly with a 1948 prediction by Ralph Alpher and Robert Hermann that the theorized 'Big Bang' origin of the Universe would leave behind an afterglow of primordial light from the time when the opaque fireball of the infant Universe became transparent (see page 166). After billions of years of travel through space, this light is finally reaching Earth, but it has been red shifted

A detailed map of the CMBR from NASA's Wilkinson Microwave Anisotropy Probe (WMAP) spacecraft combines the results from nine years of observations. Lighter areas are very slightly warmer than the CMBR's average temperature of 2.73 K, while darker areas are slightly cooler.

into the microwave part of the spectrum, creating the so-called Cosmic Microwave Background Radiation (CMBR).

In the years following the initial discovery of the CMBR, astronomers refined their temperature measurements and found that it is actually a uniform 2.73 kelvin (2.73°C above absolute zero, equivalent to −270.4°C or −454.8°F). However, the apparent uniformity of the radiation became a problem in itself, since it was difficult to match with the properties of the Universe as we know them today (see page 166). In 1992, the Cosmic Background Explorer (COBE) satellite finally resolved this problem by discovering tiny variations (about one part in 100,000) in the temperature of the CMBR. These are the seeds of the large-scale structures found across the present-day cosmos. Since then, the CMBR has been measured with ever-increasing precision, becoming an important tool for understanding conditions in the immediate aftermath of the Big Bang.

The condensed idea
The further away we see, the deeper we look back in time

45 Dark matter

The idea that more than 80 per cent of all the matter in the Universe is not only dark, but simply doesn't interact with light at all, is one of the most puzzling aspects of modern cosmology. The evidence for dark matter is overwhelming, but information about its actual composition remains frustratingly elusive.

In 1933, not long after the confirmation of galaxies beyond the Milky Way and the initial recognition of galaxy clusters as physical structures (see page 156), Fritz Zwicky made the first rigorous attempt to estimate the mass of galaxies. He investigated various methods, the most intriguing of which was a mathematical technique known as the Virial theorem – a means of estimating the mass of galaxies in a cluster from their motion and position. When Zwicky applied this theorem to the well-known Coma Cluster, he found its galaxies were behaving as if they had 400 times the mass suggested by their visible light. He attributed this difference to so-called *dunkle Materie*, or dark matter.

Zwicky's idea tied in with the discoveries of Jan Oort, who had been busy closer to home measuring the rotation of the Milky Way (see page 137). Oort had found that while the speed of objects orbiting the centre of our galaxy falls away with distance (just as the more distant planets of our own solar system orbit the Sun more slowly), they do not slow down as much as one would expect if the Milky Way's mass distribution matched that of its stars. Oort therefore suggested that there was a large amount of unseen matter filling up the Milky Way's halo region, beyond the visible spiral arms.

TIMELINE

1932	1933	1975
Oort outlines problems in the rotation of stars around the Milky Way that imply missing mass	Zwicky uses the Virial theorem to weigh the Coma Cluster, and discovers large amounts of dark matter	Rubin publishes evidence for dark matter from a detailed study of galactic rotation

Despite these early investigations, the study of dark matter was derailed for several decades by advances in other fields of astronomy. The discovery of huge clouds of interstellar gas visible at radio wavelengths – many of which were mapped by Oort himself – seemed to neatly resolve the issue. Galaxies, it was proved for certain, do contain far more matter than suggested by visible light alone. As rocket-borne and satellite telescopes opened up even more of the invisible spectrum from the 1950s onwards, more of this material, from infrared dust clouds between the stars to hot X-ray gas surrounding galaxy clusters (see page 157), was brought to light.

> IN A SPIRAL GALAXY, THE RATIO OF DARK-TO-LIGHT MATTER IS ABOUT A FACTOR OF TEN. THAT'S PROBABLY A GOOD NUMBER FOR THE RATIO OF OUR IGNORANCE-TO-KNOWLEDGE.
>
> Vera Rubin

REDISCOVERING DARK MATTER

Thus, the problem lay neglected until 1975, when US astronomer Vera Rubin published the results of her painstaking new investigation into the galactic rotation problem. She found that, once all the interstellar gas and dust was taken into account, the orbits of stars were *still* not behaving as they should. Zwicky's figures were significantly out, but galaxies seemed to behave as if they weighed about six times more than the visible matter inside them.

Rubin's claims were understandably controversial, but her work was meticulous, and once it was confirmed independently in 1978, most astronomers turned their attention from the question of whether dark matter existed, to what it could possibly be and how it could be studied.

Most attempts to explain dark matter fall into two categories: either it is due to large amounts of ordinary matter that we simply don't see because it barely emits radiation (so-called baryonic dark matter), or it is due to some new and exotic form of material (non-baryonic dark matter). In the 1980s, researchers coined catchy acronyms for the two most likely candidates – baryonic MACHOs and non-baryonic WIMPs.

1998

Japanese researchers confirm that neutrinos have mass, accounting for a small fraction of dark matter

2003

Richard Massey *et al.* use gravitational lensing to measure the distribution of dark matter in the so-called Bullet Cluster

Dark matter and the Big Bang

One other important line of evidence points to the existence of non-baryonic dark matter – the Big Bang theory itself. Not only does the Big Bang nucleosynthesis model for the creation of the elements neatly match the proportions of baryonic matter seen in the early Universe (leaving no room for MACHOs), but some form of WIMP particles are needed to explain the formation of structure in the Universe itself. The small-scale variations in the Cosmic Microwave Background Radiation (see page 178) suggest concentrations of matter and mass had already begun to form in the very early Universe, well before it became transparent. Interactions with light would have created radiation pressure that prevented baryonic matter from coalescing until after the initial fireball cleared (see page 166). Fortunately, dark matter was already able to begin building the framework around which galaxy superclusters later formed.

MACHOs (massive compact halo objects) are small but dense accumulations of normal matter thought to orbit in galaxy halos. These might include hypothetical stray planets, black holes, dead neutron stars and cooled-down white dwarfs. Such objects could have gone largely undetected by older telescopes and might account for a large amount of mass. However, improvements in telescope technology and ingenious new techniques allowed for intensive surveys of the galactic halo region in the 1990s. While some stray objects were detected, researchers concluded that they simply don't exist in the numbers needed to make a substantial contribution to dark matter.

THE QUEST FOR WIMPS

With MACHOs dismissed, astronomers and cosmologists are left with the unsettling idea of exotic WIMPs – ghostly material that somehow exists in conjunction with everyday baryonic matter, but rarely interacts with it. WIMP particles don't absorb, scatter or emit light, and may pass straight through atoms of normal matter as if they didn't exist. The only way of 'seeing' them is through the effects of their gravity on other objects.

An important first step to understanding WIMPs is to measure their distribution in relation to normal matter: are they 'cold', hanging around in close association with luminous objects, or 'hot', flying off across large distances and maintaining only the loosest connection to the visible Universe? Since the 1990s, astronomers have been developing a new technique to weigh and even map dark matter using gravitational lensing, the way in which large concentrations of mass such as galaxy clusters bend and distort the light from more distant objects (a consequence of general relativity, see page 190). Ironically, Zwicky was advocating the use of gravitational lensing to weigh galaxies as early as 1937, more than 40 years before the first examples of such objects were discovered.

By comparing the strength of lensing effects (dominated by dark matter) with the light from visible matter, researchers have discovered that the two tend to have similar distributions, suggesting that cold dark matter is the dominant type. Hot dark matter including neutrinos (the only form of WIMP to be experimentally discovered so far – see box) makes a relatively minor contribution. Yet despite these successes, the very nature of the dark matter mystery makes it more likely to yield its resolution through research at particle accelerators like the Large Hadron Collider, than through telescopic observations.

The neutrino contribution

The properties of hypothetical WIMPs match very well with those of neutrinos, the apparently massless particles emitted during certain nuclear reactions, which astronomers use to probe the interiors of stars and as an early warning of incipient supernovae (see page 122). Neutrinos are best observed using detectors deep underground, which rely on rare interactions between neutrinos and baryonic matter that produce a measurable outcome, such as a faint flash of light. In 1998, researchers at Japan's Super-Kamiokande neutrino observatory used this technique to identify a phenomenon called oscillation, in which neutrinos vary between three different 'flavours'. According to particle physics, this can only happen if the neutrinos do in fact carry a small amount of mass, though it's probably less than 1 billionth that of a hydrogen atom.

The condensed idea
Eighty per cent of the mass in the Universe is made up of mysterious invisible matter

46 Dark energy

The discovery that the expansion of the Universe is accelerating, rather than slowing down, is one of the most exciting scientific breakthroughs of recent times. Astronomers are still uncertain over exactly what dark energy is, but the possible solutions have huge implications for our understanding of the cosmos.

When NASA launched the Hubble Space Telescope in April 1990, its primary, or 'Key Project', was to establish the Hubble Constant (the rate of cosmic expansion) and therefore the age of the Universe, extending the reliable use of Cepheid variable stars as standard candles (see page 146) to unprecedented distances. This ultimately yielded a widely accepted age for the Universe of around 13.8 billion years.

SUPERNOVA COSMOLOGY

In the mid-1990s, two separate teams pioneered a new technique for measuring intergalactic distances, using Type Ia supernovae as 'standard candles', with the aim of crosschecking Hubble's results. In theory, these rare events – triggered when a white dwarf in a close binary system exceeds the Chandrasekhar limit and destroys itself in a blast of energy (see page 130) – always release the same amount of energy and should always show the same peak luminosity. The maximum brightness as seen from the Earth therefore easily yields the supernova's distance. The main challenge is that these events are extremely rare, but both teams were able to use automated search technology to scan a host of distant galaxies for the telltale early signs of brightening and catch them before they hit maximum.

TIMELINE

1915	1929	1998
Einstein adds a 'cosmological constant' term in general relativity in order to keep the Universe static	The discovery of cosmic expansion appears to render the cosmological constant redundant	Two teams of astronomers report evidence that the rate of cosmic expansion is accelerating

The idea of both projects – the international 'High-Z Supernova Search Team' and the 'Supernova Cosmology Project' based in California – was to compare the independent distances from the supernova measurements with those implied by Hubble's Law (see page 161). In total, the teams harvested data for 42 high-red-shift supernovae with distances of several billion light years, and 18 more in the nearby Universe. Because their measurements stretched far beyond the relatively local scale of the Hubble Key Project, the astronomers expected to find evidence that cosmic expansion had slowed slightly since the Big Bang. In this case, the actual distance to the furthest supernovae would be less than that suggested by their red shift, and they would therefore appear brighter than predicted.

What no one expected was that the reverse was true. The most distant supernovae seemed to be consistently *fainter* than their red shift predicted. The astronomers spent months investigating possible causes for the difference, before presenting their findings to the wider community in 1998. The inescapable conclusion was that when all other factors are taken into account, distant Type Ia supernovae really are fainter than predicted, implying that cosmic expansion has not slowed down over time, but accelerated. This surprising result has now been supported by evidence from several other approaches, including measurements of detail in the Cosmic Microwave Background Radiation (CMBR) and studies of large-scale cosmic structure. The term 'dark energy' was coined in 1998, and in 2011, Saul Perlmutter of the Supernova Cosmology Project shared the Nobel Prize in Physics

In 1994, the Hubble Space Telescope imaged a Type Ia supernova (lower left) in the relatively nearby galaxy NGC 4526. At a distance of 50 million light years from Earth, it was too close to be affected by dark energy.

ASTRONOMERS OUGHT TO BE ABLE TO ASK FUNDAMENTAL QUESTIONS WITHOUT [PARTICLE] ACCELERATORS.
Saul Perlmutter

1998
Michael Turner coins the term 'dark energy' to describe the mysterious cosmic acceleration.

2011
Perlmutter, Schmidt and Riess are awarded the Nobel Prize in Physics

2013
Planck data shows that dark energy accounts for 68.3 per cent of all energy in the Universe

Chasing vacuum energy

If dark energy is indeed best explained by a 'cosmological constant' energy field pervading space, then it may help resolve a century-old problem known as the vacuum catastrophe. Quantum theory (the physics of the subatomic world, in which waves and particles are interchangeable and familiar certainties are replaced with probabilities) predicts that any region of empty space nevertheless contains a 'vacuum energy'. This allows it to spontaneously create 'virtual' particle–antiparticle pairs for a brief moment.

The strength of this energy can be predicted from well-known principles of quantum physics, and the presence of virtual particles, popping in and out of existence all around us all the time, can be proved and even measured by a strange phenomenon called the Casimir effect. However, the measured values of vacuum energy are at least 10^{100} times weaker than the predicted effect (that's a 1 followed by 100 zeros). Little wonder, then, that vacuum energy has been called the worst theoretical prediction in the history of physics.

At first glance vacuum energy sounds a lot like the 'cosmological constant' approach to dark energy, and it would be surprising to find the two phenomena were independent. But if so, then dark energy would only make the situation worse: according to best estimates it is 10^{120} times too weak to match the predictions!

with Brian Schmidt and Adam Riess of the High-Z Supernova Search Team.

THE NATURE OF DARK ENERGY

So what exactly *is* dark energy? Various interpretations have been put forward, and just about the only thing everyone can agree on is that, in terms of energy, it's the major component of today's Universe. In 2013, measurements of the CMBR by the European Space Agency's Planck satellite suggested that dark energy accounts for 68.3 per cent of all energy in the cosmos, with dark matter making up 26.8 per cent, and ordinary baryonic matter relegated to a mere 4.9 per cent. Continuing investigation of high-red-shift supernovae, meanwhile, has further complicated matters by showing that expansion was indeed decelerating as expected in the early stages of cosmic history, only for dark energy to exerted its influence in the past 7 billion years and cause the rate of expansion to accelerate.

On hearing of the new discovery, many astronomers were reminded of Einstein's cosmological constant. The great physicist added this additional term to his general theory of relativity in order to prevent the Universe (then thought to be static) from collapsing in on itself (see page 161), but Einstein came to regret including it when cosmic expansion was later confirmed. Nevertheless, a modified version of Einstein's concept is one of the two plausible candidates for dark energy. In this model, the constant is a tiny amount of energy intrinsic to a fixed volume of space. Since energy is equivalent to mass through the famous equation $E = mc^2$, the constant therefore has a gravitational effect just like any mass, though for complex reasons the effect is repulsive in this case. Despite the energy content of each cubic kilometre of space being tiny, the effects mount up over large distances. They also increase over time as the Universe, and the volume of space within it, expand.

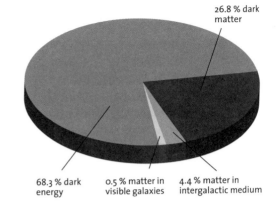

26.8 % dark matter

68.3 % dark energy

0.5 % matter in visible galaxies

4.4 % matter in intergalactic medium

This pie chart shows the dominance of dark energy in the Universe's mass-energy content, according to 2013 measurements from the Planck telescope.

The major alternative candidate explanations are so-called 'quintessence' theories, in which the energy density of dark energy is not uniform across space, but is instead dynamic, accumulating in some places more than others and causing them to expand more. Various theories of this kind have been put forward, some of which treat quintessence as a 'fifth force' of nature to rank alongside gravity, electromagnetism and the forces of the atomic nucleus.

Whatever the true nature of dark energy, scientists will continue to study its effects in today's Universe and in the past. The implications for the future of the cosmos, meanwhile, are huge, and potentially doom it to a long, cold death (see page 200).

The condensed idea
Cosmic expansion is accelerating, but we're not sure why

47 Relativity and gravitational waves

Einstein's twin theories of special and general relativity revolutionized physics in the early 20th century. For theoretical cosmologists, they provide the foundations for understanding the nature of the Universe, while for working astronomers, they can offer new tools for observing the extremes of the cosmos.

Albert Einstein was a failed academic working at the Swiss Patent Office when in 1905 he published a series of four papers that propelled him to scientific fame. Two of these were concerned with the atomic and subatomic realm, but the second pair investigated the behaviour of objects in non-accelerating motion at close to the speed of light – the phenomenon known as special relativity. Einstein was driven to investigate the extremes of motion by problems that had manifested in physics over the previous decade – in particular, questions around the speed of light.

Scots physicist James Clerk Maxwell had established in 1865 that light has a fixed speed in vacuum (denoted c) of around 300,000 kilometres per second (186,000 miles per second). Physicists assumed at the time that this was its speed of transmission through an all-pervading, light-transmitting medium they called the luminiferous aether. With sensitive measurement techniques, it should be possible to measure the slight variation in the speed of light from different directions caused by Earth's motion through the aether. In

TIMELINE

1865	1887	1905	1907
Maxwell calculates the fixed speed of light and other electromagnetic radiation in a vacuum	Failure of the Michelson–Morley experiment throws physics into crisis	Einstein publishes his theory of special relativity, including the equivalence of mass and energy	Minkowski shows how special relativity can be treated as a geometric effect in four-dimensional spacetime

1887, therefore, Albert Michelson and Edward Morley devised an ingenious and highly sensitive new experiment to detect this difference,. When their experiment drew a blank, physics was thrown into something of a crisis. Various theories were put forward to explain away the negative result, but only Einstein dared to take it at face value, and consider the possibility that the aether really didn't exist. Instead, he wondered, what if the speed of light is simply a constant, regardless of the relative motion of source and observer?

SPECIAL RELATIVITY

Einstein's first paper essentially reimagines the simple laws of mechanics based on two axioms: the fixed speed of light, and the 'principle of relativity' (i.e. that the laws of physics should always appear the same to observers in different but equivalent frames of reference). Putting situations of acceleration to one side, he considered only the 'special' case of inertial (non-accelerating) frames of reference. In most everyday situations, he showed, the laws of physics will be the same as those outlined by Isaac Newton in the late 17th century. But when observers in two different reference frames are in relative motion at close to the speed of light, they start to interpret events in radically different ways. These so-called 'relativistic' effects include a contraction of length in the direction of motion, and a slowing down of time (time dilation). In his second paper on relativity, Einstein showed that objects travelling at relativistic speeds also increase in mass, and from this, he proved that a body's mass and energy are equivalent, deriving the famous equation $E = mc^2$. In all cases, the distortions are only apparent to an observer *outside* the frame of reference. For anyone within it, everything appears normal.

> THE QUEST OF THE ABSOLUTE LEADS INTO THE FOUR-DIMENSIONAL WORLD.
>
> Arthur Eddington

Special relativity is important to astronomers because it implies there is no fixed frame of reference from which the Universe should be measured – nowhere that is truly stationary or where time runs at an absolute speed. The effects it

1915

Einstein publishes his theory of general relativity, showing how mass warps spacetime

1919

Eddington demonstrates the gravitational lensing effect arising from general relativity

2016

LIGO scientists confirm the existence of gravitational waves, the last unproved prediction of general relativity

Gravitational lensing

Gravitational lensing occurs when light rays from distant objects pass close to a large mass and have their paths deflected by the distorted spacetime around it. For an object like the Sun, the effect is barely detectable (Eddington's eclipse expedition measured deflections in the apparent position of stars amounting to less than one ten-thousandth of a degree), but for larger concentrations of mass the results can be far more impressive. The first such gravitationally lensed object, discovered in 1979, is the Twin Quasar – a distant quasar whose light reaches Earth from two directions after deflection around an intervening galaxy.

Astronomers have since found many more examples of gravitational lensing – particularly around dense galaxy clusters, where light from the lensed background object is often distorted into a series of arclike patterns. Lensing offers a powerful tool for mapping the distribution of mass within such clusters to learn more about the presence of dark matter (see page 160), but it can also have a more direct application. Just like a glass telescope lens, a gravitational lens can intensify the light of more distant objects, bringing extremely faint galaxies at the limit of visibility within range of powerful telescopes. Indeed, this is how astronomers detected the most distant galaxy yet discovered, some 13.2 billion light years away.

predicts have not only been demonstrated in Earth-based experiments, but have also proved useful in explaining a variety of astronomical phenomena, ranging from the behaviour of relativistic jets (emitted from the poles of neutron stars and active galaxies) to the origin of matter in the Big Bang itself.

FROM SPECIAL TO GENERAL

In 1915, Einstein published a more general theory, now incorporating situations in which acceleration was involved. The key breakthrough came in 1907, when he realized that, since gravity causes acceleration, a person in a situation of steady acceleration should observe exactly the same laws of physics as someone standing on the surface of a planet in a gravitational field. For astronomers, this has an important implication: just as an observer moving in a rapidly accelerating rocket sees the path taken by a beam of light curve downwards, so the same thing should happen in a strong gravitational field. This is the root of the spectacular phenomenon known as gravitational lensing (see box, left).

Over the next eight years, Einstein laboured over the implications of his discovery, heavily influenced by his former University tutor Hermann Minkowski's ideas about special relativity. Minkowski had explored relativistic distortions through the rules of geometry, treating the three dimensions of space and one of time as a unified structure or spacetime manifold, within which each dimension can be traded off against the others. Einstein imagined gravity as a distortion to spacetime, and developed equations to describe it.

His 1915 paper applied the new theory to explain features of the planet Mercury's orbit that were inexplicable in classical physics, but, because it was published in German at the height of the First World War, it went largely unnoticed. It was only in 1919 that Arthur Eddington delivered a spectacular demonstration of the new theory, measuring the gravitational lensing effect on stars close to the Sun during a solar eclipse.

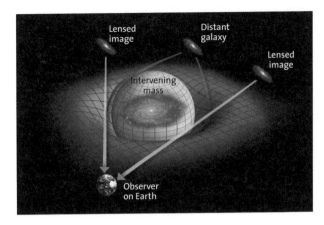

GRAVITATIONAL WAVES

In the 20th century, special and general relativity were shown to be correct again and again, but until very recently, one key prediction remained unproven. Gravitational waves are tiny ripples in spacetime, manifested as atomic-scale changes to the dimensions of space, and created by rapidly rotating nonsymmetric masses (such as black holes or neutron stars spiralling together in binary systems – see page 131).

In February 2016, however, US scientists finally announced the detection of ripples in spacetime from a pair of merging black holes, using the Laser Interferometer Gravitational-Wave Observatory (LIGO) instruments in Washington State and Louisiana. The discovery not only confirms Einstein's theory (and, indeed, proves the existence of black holes beyond doubt), but also opens up a powerful new method for observing the cosmos. Since gravitational waves are created by mass rather than luminous matter, future gravitational wave observatories should be able to study dark matter and even peer beyond the limits of the decoupling era (see page 166) to study conditions in the Big Bang itself.

One popular way of thinking about general relativity is to imagine spacetime as a rubber sheet, within which massive objects create distortions. These not only affect the orbits of other objects, but also deflect the path of light, giving rise to the phenomenon known as gravitational lensing.

The condensed idea
Space and time are intertwined

48 Life in the Universe

The search for extraterrestrial life and intelligence is one of the most challenging but exciting areas of modern astronomy. Yet, even without further discoveries, the existence of our own habitable planet raises an intriguing question: why should the Universe be able to support life at all?

The past few decades have seen a revolution in the prospects for life in our galaxy and the wider Universe (see Chapters 12 and 26). But the bigger question is one of intelligence: proof of extraterrestrial life would forever change our understanding of the cosmos, but contact from an alien species with its own science, technology and philosophy would be a far more profound and transformative event.

SIGNAL HUNTING

Various projects aimed at detecting signs of alien life have been organized since the early 1960s. Collectively termed the Search for ExtraTerrestrial Intelligence (SETI), they generally focus on surveying the sky at radio wavelengths, looking for signals that cannot be explained through natural phenomena. Although this approach may be the best one we have, there are obvious drawbacks: radio signals, like all electromagnetic waves, spread out and fade rapidly unless they are collimated in a tight directional beam, meaning we are essentially relying on a 'communicating civilization' deliberately sending a signal towards our small region of space. This may not be as unlikely as it seems, since we might do something very similar if we ever detected signs of life on an alien exoplanet.

TIMELINE

1960	1961	1973
Frank Drake uses the Green Bank radio telescope in the first modern SETI research	Drake formulates an equation for finding the number of civilizations in our galaxy, though it contains many unknown factors	Carter uses his anthropic principles to explain why the Universe is hospitable to life

More problematically, any alien broadcasters would have to keep their radio antenna pointing in our direction for a long period of time, as the chances of us looking in the right direction at the right moment, with our telescopes tuned to the right frequency, would be astronomically small. Even if such a lucky coincidence occurred, it might easily be dismissed as a rogue 'one-off' unless the signal was repeated. The most exciting candidate for an extraterrestrial radio signal to date – the so-called 'Wow!' signal detected by SETI scientist Jerry Ehman in August 1977 – failed on this criterion. This burst of radio waves, apparently emanating from Sagittarius, has never been repeated despite numerous searches.

In light of these problems for the traditional radio approach, some SETI astronomers have pioneered alternative ideas. Optical SETI advocates search for signals being sent by means of visible light, while 'Active SETI' supporters have sent deliberate messages into space, most famously the 1974 Arecibo message (see illustration).

Another promising approach is to look for 'technosignatures' rather than deliberate messages. These are telltale oddities in the light of stars and planets that could only be created by the activities of an advanced civilization. It sounds like science fiction at first, but structures such as planetwide cities, star-shifting 'Shkadov thrusters' and Dyson spheres (huge shells constructed around a star in order to harvest its energy) are all plausible engineering projects that would produce a distinctive signal. What's more, this approach has already yielded the

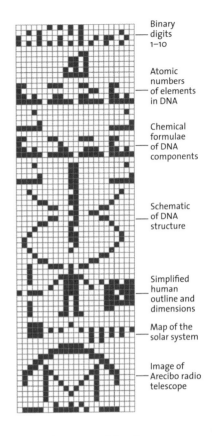

Binary digits 1–10

Atomic numbers of elements in DNA

Chemical formulae of DNA components

Schematic of DNA structure

Simplified human outline and dimensions

Map of the solar system

Image of Arecibo radio telescope

The Arecibo Message (above) was a burst of 1679 binary digits – the product of the prime numbers 23 and 73. When laid out on a grid of 23 columns and 73 rows, the message forms a simple pictogram.

1974

Drake, Carl Sagan and others collaborate to send the symbolic Arecibo message to a distant star cluster

1977

Ehman detects a strong, non-repeating radio signal apparently originating from the direction of Sagittarius

1986

Barrow and Tipler reformulate the weak and strong anthropic principles in their modern form

The mystery of Tabby's Star

In September 2015, a team of astronomers led by Tabetha Boyajian at Yale University announced their discovery of inexplicable variations in the light of a star called KIC 8462852 (subsequently nicknamed Tabby's Star). Lying about 1,480 light years away in the constellation of Cygnus, the star had been discovered as part of the Kepler search for exoplanets (see page 99), but the intermittent dips in its light cannot be explained by transiting planets. Instead, they appear to indicate a swarm of smaller bodies in orbit. The most plausible natural explanation therefore seemed to be a large number of comets on highly elliptical paths that just happened to cross in front of the star during Kepler's observations, but as SETI researcher Jason Wright pointed out, the dips could also be caused by a Dyson sphere or similar structure being assembled in orbit. Radio surveys failed to detect any unusual signals coming from the star's vicinity, but the mystery deepened in early 2016 when astronomer Bradley Schaefer checked historical records and found that KIC 8462852 has faded in brightness by about 20 per cent since 1890, more or less ruling out the comet explanation. Tabby's Star probably isn't home to an alien construction site, but it certainly lives up to its billing as the 'most mysterious star in our galaxy'.

most exciting SETI candidate in years (see box on page 194).

FINE-TUNED FOR LIFE?

While SETI astronomers have been busy looking for intelligence, some cosmologists have been equally absorbed by the question of why *any* planet in the Universe should have life – and indeed, why there are any planets, stars or galaxies at all. As the Big Bang theory has developed since the 1950s, it's become increasingly clear that many aspects of our Universe, from the large-scale structure of galaxy clusters and superclusters to the chemical behaviour of individual elements, are dependent on a handful of physical constants. If any of these had a slightly different value, then the Universe as a whole would be very different – probably different enough to preclude the development of life. Given that the Big Bang theory itself offers no specific mechanism for controlling the values of these constants, the fact that they seem to be 'fine-tuned' for life looks like an extraordinary coincidence.

Physicist Robert Dicke was the first to argue a possible explanation for this fine-tuning in 1961, when he noted that we can only exist because we live at a specific stage in cosmic history that is suitable for the evolution of life. Therefore, we should not affect to be surprised at the fact we live at that particularly hospitable time. The same basic idea lay at the heart of the 'anthropic principle', proposed in two forms by US astrophysicist Brandon Carter in 1973. Carter's weak anthropic principle states simply that because we are here, our location in space and time *must* be one suited for giving rise to life. His strong principle makes the same argument for the values of the physical constants, pointing out that if they were wildly different, we wouldn't be around to measure them.

In 1986, cosmologists John Barrow and Frank Tipler revisited the question in a bestselling book, *The Anthropic Cosmological Principle*. Confusingly, they came up with their own definitions of the weak and strong principles that were rather different from Carter's, and it is these versions that are generally used in discussion today. Barrow and Tipler's weak anthropic principle essentially encompasses both Carter's weak and strong variants, arguing that all physical aspects of the Universe will naturally be suited to life simply because we are here to measure them. Their strong principle, however, goes much further, suggesting there may be something about the Universe that gives it an *imperative* to produce life – in other words, it really *is* fine-tuned by an outside influence. The authors put forward three possible explanations for their strong principle: either the Universe was deliberately designed to give rise to life by an external agency; or the presence of observers is in some way necessary for the Universe to come into being (an approach that echoes some elements of quantum physics); or finally, our Universe is just one of many in a vast 'ensemble' that allows all the possible parameters to be explored. As we'll see in Chapter 49, this third option may not be as unlikely as it sounds.

> **TWO POSSIBILITIES EXIST: EITHER WE ARE ALONE IN THE UNIVERSE OR WE ARE NOT. BOTH ARE EQUALLY TERRIFYING.**
> Arthur C. Clarke

The condensed idea
The Universe's suitability for life raises awkward questions

49 The multiverse

Could our Universe be just one tiny part of a much larger and perhaps infinite multiverse? Many cosmologists are increasingly keen on the idea, but what evidence might be found to support it? And what form would the unseen parts of the multiverse take?

Probably the most widely known form of multiverse is also the hardest to imagine. This is the ensemble of infinite parallel universes suggested by the 'many worlds interpretation' of quantum mechanics, and beloved of science-fiction writers. According to this idea, first put forward by physicist Hugh Everett in 1957, the solution to the uncertain outcomes inherent in the subatomic world of quantum theory is for the Universe to constantly branch, spawning copies in which every possible outcome of every possible event is played out. Fortunately, the two kinds of multiverse most commonly advocated by cosmologists are somewhat easier to grasp, though their implications are in many ways just as profound.

BEYOND THE LIMITS

The simplest object that might be called a multiverse is one that we can be certain exists – the extension of our own cosmos far beyond the 46.5-billion-light-year limit of the 'observable Universe' set by the speed of light (see page 177). The existence of such a multiverse is fairly obvious when you consider the situation for a hypothetical observer on a planet at the edge of *our* observable Universe. Looking in one direction they see across the gulf of space towards the Earth, but looking in the other they can view regions

TIMELINE

1957	1981	1983
Hugh Everett formulates the many worlds interpretation of quantum mechanics	Alan Guth suggests our Universe is just a small inflated bubble of the original Big Bang	Steinhardt argues that inflation may be an eternal process

of spacetime forever off-limits to our own observations.

Based on the evidence that our visible Universe is 'homogenous and isotropic' on the largest scales (in other words, it looks roughly the same regardless of where you are or in which direction you're looking), it's reasonable to conclude that this multiverse is essentially 'more of the same', but just how big is it? The answer to this question depends on the curvature of spacetime itself, determined by the balance of matter, dark matter and dark energy in the cosmos (see page 200). If spacetime curves inwards like a sphere, then the multiverse is closed, and perhaps no more than 250 times larger than our visible Universe. If spacetime flexes outwards like a saddle, however (as the discovery of dark energy suggests), then the multiverse is open and effectively infinite in size. Strangely, a truly infinite universe carries with it the same implication as the many worlds hypothesis – somewhere out there, every possible outcome of every event is being played out in a 'parallel' Universe.

Four flavours of multiverse?

The theorist and multiverse pioneer Max Tegmark defines four levels of multiverse:

1. Normal spacetime beyond the limits of the observable Universe.
2. Universes with different physical constants, such as those created by eternal inflation.
3. The parallel Universes generated by the many worlds interpretation of quantum mechanics.
4. The ultimate ensemble – a purely mathematical structure that incorporates all possible multiverses.

ETERNAL INFLATION

The second type of multiverse that intrigues cosmologists is even more bizarre, offering the possibility of universes radically different from our own. It has its roots in the theory of inflation, devised by Alan Guth and others in the early 1980s as a means of blowing up a small section of the primordial cosmos and creating a Universe like the one we see today (see page 166). An

1986

Andrei Linde proposes a chaotic inflation model that produces an infinite number of bubble universes

1995

Edward Witten develops brane theory as a variation of string theory

2001

Steinhardt and Turok publish their brane cosmology theory of the multiverse

Branes and higher dimensions

Attempts to find a unifying theory of particle physics in the past few decades have given rise to another possible form of multiverse, known as brane cosmology. The most likely current candidate for uniting the fundamental forces of nature – a complex idea known as M-theory – requires that spacetime should contain seven extra space dimensions, of which we are currently unaware. Some of these could be 'compactified', or curled in on themselves at such small scales that they go unnoticed in our Universe (in the same way that a small ball of string looks like a single point if viewed from far enough away), but what if one of them was not?

In the late 1990s, cosmologists developed a theory that our Universe might just be a membrane-like region of spacetime called a brane, separated from a multiverse of similar branes by small distances in an unseen 'hyperspace' dimension. In 2001, Paul Steinhardt and Neil Turok used branes as the basis for a new cyclic model of cosmic evolution, suggesting that branes move slowly apart in hyperspace, and this manifests itself as dark energy within each brane. Collisions between branes on trillion-year timescales trigger Big Crunch events (see page 201), followed by new Big Bangs.

obvious question at the time was what caused inflation to come to an end but, in 1986, Guth's sometime collaborator Andrei Linde raised a more daring possibility – what if inflation *never* came to an end?

In the eternal or chaotic inflation model, new 'bubble Universes' are continuously being created by a process of phase change, which is analogous to the formation of bubbles in soda water. In everyday life we're familiar with the solid, liquid and gaseous phases of materials, and perhaps vaguely aware that transitions between them absorb or release energy. But in fundamental physics, many more properties have phases, ranging from the characteristics of elementary particles to the dimensions of spacetime itself, and the vacuum energy that permeates the cosmos (see page 186).

Transitions between these phases release far more energy than those between the phases of matter, and new phases can pop into existence spontaneously in the vacuum of space. Their fate then depends on their precise mix of properties – those with negative vacuum energy rapidly collapse back on themselves, but those with positive energy start to expand, potentially creating a bubble Universe with its own properties and physical laws, and even its own mix of dimensions. In many cases, the vacuum energy could be much larger than it is in our own Universe, perhaps driving a universe that expands exponentially. Beyond our particular bubble, the wider multiverse would be anything but homogenous and isotropic.

INFINITE VARIETY

If this model of the multiverse is correct, then it solves many of the mysteries of modern cosmology. For instance, the existence of countless phases with radically different properties would render the fine-tuned nature of our own Universe and the low value of its vacuum energy less problematic (see pages 194 and 186). The question of what happened 'before' the Big Bang and how it was triggered would finally become meaningful, but conversely, our long-held picture of a 13.8-billion-year-old Universe would need to be abandoned, since this would only be the age of our particular bubble in an eternal process.

For the moment, however, this extraordinary theory remains unproven. Some might ask whether it would ever be possible to confirm the existence of such varied universes beyond our own, but one advantage of eternal inflation is that it does make testable predictions. In theory, bubbles should occasionally impinge on each other, with their outer walls smashing together at high speeds. The result in our Universe would be a 'cosmic wake' whose passage would have various effects, and which would imprint distinctive patterns onto the Cosmic Microwave Background Radiation. Although such patterns have not yet been found, they would be at the very limits of present-day observing techniques, so the case for this kind of multiverse remains tantalizingly unproven.

> **IN INFINITE SPACE, EVEN THE MOST UNLIKELY EVENTS MUST TAKE PLACE SOMEWHERE.**
> Max Tegmark

The condensed idea
Our Universe may be just one of many in an infinite cosmos

50 Fate of the Universe

Just what is the ultimate fate of our Universe? Since the birth of modern cosmology, astronomers have sought to distinguish between several distinct alternatives, but the recent discovery of dark energy has introduced an important new factor, apparently dooming the cosmos to a long, cold death.

The idea that the Universe might one day come to an end was as alien to astronomers in the early 20th century as the idea that it had a beginning. Up to this point, the cosmos had been generally considered to be eternal, and to have been much the same in the distant past as it is today. The first person to seriously consider the alternative was Alexander Friedmann, who in 1924 built on his earlier idea of expanding spacetime (see page 161) with thoughts on how the Universe might evolve. Friedmann argued that the Universe must be expanding in order to overcome the gravitational influence of the matter within it. How long this expansion would continue depended on a crucial factor known as the density parameter (denoted by the Greek letter omega, Ω) – the average distribution of mass and energy compared to a certain critical density.

If Ω is exactly 1 (i.e. the average density of the Universe is equal to the critical density), then gravity will be sufficient to slow down cosmic expansion, but never quite bring it to a halt. If Ω is less than 1, then expansion will continue forever, while if it is more than 1, it will slow down and eventually reverse, with the entire Universe falling back upon itself. Friedmann described these three scenarios as flat, open or closed, respectively.

TIMELINE

1924	1934	1969	1977
Friedmann studies the possible expansion of spacetime	Tolman shows that an oscillating universe breaches the laws of thermodynamics	Rees considers conditions in a closed 'Big Crunch' universe	Islam studies the long-term fate of matter in an open universe

Following Friedmann's work and Hubble's 1929 confirmation of cosmic expansion, Einstein, Lemaître and others considered the possibility of a cyclic or oscillating Universe that periodically expanded and contracted, going through a hot, dense state at either end of the cycle (a Big Bang and a Big Crunch). A cyclic Universe seemed more eternal than the definitive moment of creation suggested by a straightforward expansion model, but in 1934 Richard Tolman showed that no oscillating universe could continue forever without breaching the laws of thermodynamics. It would still need a definitive beginning, so advocates were just swapping a more recent moment of creation for a distant one.

> **THE LAWS OF NATURE ARE CONSTRUCTED IN SUCH A WAY AS TO MAKE THE UNIVERSE AS INTERESTING AS POSSIBLE.**
> Freeman Dyson

BIG CRUNCH OR HEAT DEATH?

After this early flurry of interest, the future evolution of the Universe remained something of a scientific backwater until the mid-1960s, when the Big Bang theory was conclusively proved by the discovery of the Cosmic Microwave Background Radiation. In 1969, Martin Rees revisited the subject with a review of conditions in a collapsing closed Universe. He found that as the Universe contracted, it also heated up, eventually reaching temperatures that would cause stars themselves to evaporate, before everything was either destroyed in a singularity, or recycled in an oscillating Universe.

In 1977, meanwhile, Bangladeshi cosmologist Jamal Nazrul Islam made the first study of what might happen in an open Universe. He predicted that, over trillions of years and even longer, much of the material in galaxies would ultimately end up as collapsed black holes that slowly radiate away their mass through Hawking radiation (see page 134). Furthermore, on even longer timescales, many of the subatomic particles in ordinary matter would prove vulnerable to radioactive decay. Another way of looking at this scenario is through the laws of thermodynamics, as William Thomson (Lord Kelvin),

1998	2001	2002
Cosmologists discover the dark energy of cosmic expansion, suggesting the Universe must be open and unending	Steinhardt and Turok revive the idea of a cyclic universe with their brane cosmology theory	Linde argues that dark energy might be able to reverse itself in the future

A Big Slurp?

Since the 1970s, particle physicists have been aware of a possible fate for the Universe that tends to be ignored in cosmological discussions to the question. This is the possibility that the present-day vacuum of space is not as stable as it appears to be, but is instead 'metastable' and vulnerable to a possible dramatic change at some point. In physics, a metastable state is one that appears to have a minimum of energy and will be stable in most situations, but which can suddenly collapse if the possibility of dropping to an even lower energy state is introduced. On a cosmic scale, such an event could happen if a small bubble of the true vacuum state popped briefly into existence due to quantum effects (in a similar way to virtual particles – see page 186). The bubble would expand at the speed of light, destroying any matter in its path by loosening the bonds of the fundamental forces – a cataclysm nicknamed the Big Slurp. Although such an event would almost certainly be tens of billions of years in the future, calculations based on data such as the mass of the newly discovered Higgs Boson increasingly point to the idea that our Universe is indeed in a fragile metastable state.

did in the 1850s. In effect, energy and information get more and more spread out until the Universe is effectively uniform, a condition known as heat death. In 1979, Freeman Dyson treated all of these concepts in more detail in his highly influential study, *Time Without End*, laying out a scenario generally known as the Big Chill.

Distinguishing between the two scenarios of an open and closed Universe became a major preoccupation for cosmologists in the 1980s, made more difficult by the need to accurately measure the contribution of dark matter. Most estimates suggested that the Universe was hovering close to critical density, which led to a redoubling of efforts.

However, the discovery of dark energy in 1998 changed everything. The fact that cosmic acceleration is actually increasing seemed to rule out the closed and flat spacetime scenarios. In their place, the Big Chill was joined by a more alarming option. So far, So far, we don't understand enough about dark energy to know how it will behave in the future, but one possibility (dubbed phantom energy by Robert Caldwell in 2003) is that the strength of dark energy will continue to increase exponentially, eventually becoming strong enough to affect the smallest scales and tearing matter to pieces in a 'Big Rip'. Andrei Linde suggested in 2002 that it might prove capable of reversing itself, sending the Universe hurtling back towards a Big Crunch after all. Confirmation that dark energy seems to have altered its behaviour over time (see page 186) only adds to the doubt surrounding any predictions of its future strength.

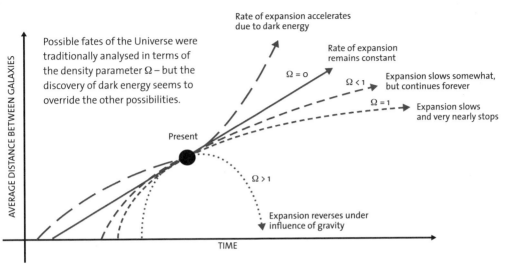

Possible fates of the Universe were traditionally analysed in terms of the density parameter Ω – but the discovery of dark energy seems to override the other possibilities.

Rate of expansion accelerates due to dark energy

Rate of expansion remains constant

Ω = 0

Ω < 1   Expansion slows somewhat, but continues forever

Ω = 1   Expansion slows and very nearly stops

Present

Ω > 1

Expansion reverses under influence of gravity

AVERAGE DISTANCE BETWEEN GALAXIES

TIME

NOT QUITE THE END?

If the idea of a long, cold cosmic twilight or a dramatic shredding of all matter doesn't lighten the heart, then ideas about the multiverse (see page 198) at least hold out some hope for the distant future. According to the eternal inflation model, new universes are springing up all the time, and one may even pop into existence inside our own region of spacetime before the long gloom sets in. Alternatively, Paul Steinhardt and Neil Turok's cyclic Universe model may offer another way to regenerate the Universe, albeit long after everything that is interesting about our own one has withered away.

The condensed idea
How will the Universe end – indeed, will it end at all?

Glossary

Active galaxy A galaxy that emits large amounts of energy from its central regions.

Asteroid One of the countless rocky objects in the inner Solar System.

Astronomical unit A unit of measurement equivalent to Earth's average distance from the Sun – roughly 150 million kilometres (93 million miles).

Atmosphere A shell of gases held around a planet or star by its gravity.

Binary star A pair of stars in orbit around each other.

Brown dwarf A 'failed star' that lacks the mass to fuse hydrogen in its core.

Comet A chunk of rock and ice from the outer reaches of the Solar System.

Dwarf planet A planet-like object that lacks the mass to qualify as a true planet.

Eclipsing binary A binary system whose stars regularly passes in front of each other, causing a drop in overall brightness.

Electromagnetic radiation A type of energy consisting of combined electric and magnetic waves, able to propagate itself across a vacuum at the speed of light.

Flare A huge release of superheated particles above the surface of a star, caused by a short-circuit in its magnetic field.

Galaxy An independent system of stars, gas and other material with a size measured in thousands of light years.

Globular cluster A dense ball of ancient, long-lived stars, in orbit around a galaxy.

Kuiper Belt A doughnut-shaped ring of icy worlds directly beyond Neptune's orbit.

Light year The distance traveled by light (or other electromagnetic radiation) in one year, equivalent to roughly 9.5 million million kilometres (5.9 trillion miles).

Main sequence A term used to describe the longest phase in a star's life, during which it is relatively stable, and shines by fusing hydrogen into helium at its core.

Nebula A cloud of gas or dust floating in space. Nebulae are the material from which stars are born, and into which they are scattered again at the end of their lives.

Neutron star The collapsed core of a supermassive star, left behind by a supernova explosion. Many neutron stars initially behave as pulsars.

Nova A binary star system with a white dwarf robbing material from a companion star and triggering occasional explosions.

Nuclear fusion The joining-together of light atomic nuclei to make heavier ones at very high temperatures and pressures, releasing excess energy in the process. Fusion is the process by which the stars shine.

Oort Cloud A spherical shell of dormant comets, up to two light years across, surrounding the entire Solar System.

Open cluster A large group of bright young stars that have recently been born from the same star-forming nebula.

Planet A spherical world orbiting a star, with enough mass and gravity to clear the space around its orbit of other objects apart from its own moons.

Planetary nebula An expanding gas cloud formed by the expelled outer layers of a dying red giant star.

Pulsar A rapidly spinning neutron star with an intense magnetic field that channels its radiation into two narrow beams.

Red dwarf A star with considerably less mass than the Sun – small, faint, and with a low surface temperature.

Red giant A star passing through a phase of its life where its luminosity has increased hugely, causing its outer layers to expand and its surface to cool.

Relativistic jets Beams of particles moving close to the speed of light, generated around objects such as black holes.

Snowline The point in any solar system where the central star's radiation is weak enough for water ice and other volatile chemicals to survive in solid form.

Spacetime A four-dimensional 'manifold' in which the three space dimensions are interlinked with the dimension of time, giving rise to the effects of special and general relativity.

Spectroscopy The study of the distribution of colours of light from stars and other objects, revealing information such as the object's chemical composition, size and motion through space.

Standard candle Any astronomical object whose luminosity can be independently known, allowing its distance to be worked out from its apparent brightness.

Star A giant ball of gas whose centre is hot and dense enough to trigger nuclear fusion reactions that allow it to shine.

Supergiant A massive and extremely luminous star with between 10 and 70 times the mass of the Sun.

Supermassive black hole A black hole with the mass of millions of stars, believed to lie in the centre of many galaxies.

Supernova A cataclysmic explosion marking the death of a star.

Transit The passage of one celestial body across the face of another.

Variable star A star whose brightness varies, either due to interaction with another star, or because of some feature of the star itself.

White dwarf The dense, slowly cooling core left behind by the death of a star with less than eight times the Sun's mass.

Wolf–Rayet star A star with extremely high mass which develops such fierce stellar winds that it blows away its outer

Index

First published in the UK in 2016 by

Quercus Editions Ltd
Carmelite House
50 Victoria Embankment
London EC4Y 0DZ

An Hachette UK company

Design and editorial by Pikaia Imaging

Edited by Dan Green

A CIP catalogue record for this book is available
from the British Library

HB ISBN 9781784296100
EBOOK ISBN 9781784296117

10 9 8 7 6 5 4 3 2 1

Printed and bound in China

Acknowledgements:

Thanks to Paul Crowther, Matthew
Kleban, Hal Levison, Giuliana de
Toma, and the many other working
scientists who have kindly made time
to help bring me up to speed on some
of the most exciting areas of modern
astronomy in recent months.

This book would not have been
possible without the able assistance of
Tim Brown and Dan Green – thanks to
both for their superhuman efforts!

And above all, thanks to Katja for her
unfailing support.

Picture credits:

26: Robin Canup, Southwest Research
Institute; 31: NASA/JPL/Malin Space
Science Systems; 42: NASA, ESA, H.
Weaver (JHU/APL), A. Stern (SwRI),
and the HST Pluto Companion Search
Team; 51: NASA; 82: ESO; 91: ESO/I.
Crossfield; 138: User: Dbenbenn via
Wikimedia; 159: 2dF Galaxy Redshift
Survey team/http://www2.aao.gov.
au/2dFGRS/; 179: NASA/WMAP Science
Team; 185: NASA/ESA, The Hubble
Key Project Team and The High-Z
Supernova Search Team; 193: Pengo via
Wikimedia.

All other pictures by Tim Brown.